Finite Element Methods for Engineers

Finite Element Methods for Engineers

Roger T Fenner

Department of Mechanical Engineering
Imperial College of Science, Technology and Medicine
London

Imperial College Press

ICP

Published by

Imperial College Press
Sherfield Building, Imperial College
London SW7 2AZ

Distributed by

World Scientific Publishing Co. Pte. Ltd.
P O Box 128, Farrer Road, Singapore 912805
USA office: Suite 1B, 1060 Main Street, River Edge, NJ 07661
UK office: 57 Shelton Street, Covent Garden, London WC2H 9HE

British Library Cataloguing-in-Publication Data
A catalogue record for this book is available from the British Library.

First published 1975 by
THE MACMILLAN PRESS LTD
© Roger T. Fenner 1975

ISBN 1-86094-016-1

This book is printed on acid free paper.

Printed in Singapore by Uto-Print.

Contents

Preface

The advent of high-speed electronic digital computers has given tremendous impetus to all numerical methods for solving engineering problems. Finite element methods form one of the most versatile classes of such methods, and were originally developed in the field of structural analysis. They are, however, equally applicable to continuum mechanics problems in general, including those of fluid mechanics and heat transfer. While some very sophisticated finite element methods have been devised, there is a great deal of very useful analysis that can be performed with the most straightforward types, which are simple to understand and formulate.

The teaching of finite element methods has hitherto been largely confined to university postgraduate courses, particularly those concerned with structural analysis in civil or aeronautical engineering. The purpose of this book is to serve as an introduction to finite element methods applicable to a wider range of problems, particularly those encountered in mechanical engineering. The main emphasis is on the simplest methods suitable for solving two-dimensional problems. Since computer programs form an integral part of the finite element approach they are treated as such in the text. Several programs are presented and described in detail and their uses are illustrated with the aid of a number of practical case studies.

This book is based on courses given by the author to both undergraduate and postgraduate students of mechanical engineering at Imperial College. A prior knowledge of the FORTRAN computer programming language is assumed. The level of continuum mechanics, numerical analysis, matrix algebra and other mathematics employed is that normally taught in undergraduate engineering courses. The book is therefore suitable for engineering undergraduates and other students at an equivalent level. Postgraduates and practising engineers may also find it useful if they are comparatively new to finite element methods.

The author wishes to thank Miss E. A. Quin for her very skilful typing of a difficult manuscript.

Imperial College of Science and Technology,   ROGER T. FENNER
London

Notation

The mathematical symbols commonly used in the main text are defined in the following list. In some cases particular symbols have more than one meaning in different parts of the book, although this should not cause any serious ambiguity.

A	constant in Lamé equations for a thick-walled cylinder
A	a square matrix
a	radius of small hole in a flat plate
a_i, a_j, a_k	dimensions of a triangular element
a_x, a_y	semi-axes of an ellipse
a_1, a_2, a_3	constants in general boundary condition equation 2.84
B	width of a beam
B	constant in Lamé equations for a thick-walled cylinder
B	element dimension matrix
B_{rs}	coefficient of B
b	bandwidth of overall stiffness matrix
b	semi-width of a square flat plate
b_i, b_j, b_k	dimensions of a triangular element
C	torsional couple
C_p	specific heat
C_1 to C_{10}	constants in polynomial shape functions
D	flexural rigidity of a flat plate
D	element elastic property matrix
E	Young's modulus
e	strain
e	element strain vector
e_T	truncation error
e_T	element thermal strain vector
F	vector of overall externally applied forces
F_i	externally applied force at node i
F_i	subvector of externally applied force components at node i

F_m	element vector of externally applied forces
F_D, F_P	drag and pressure flow shape factors for downstream flow
f_i	self-flexibility submatrix for node i
f	coefficient of f
G	shear modulus
G	vector of overall body forces applied to the nodes
G_i	subvector of overall body force components at node i
G_m	element body force vector
g	heat generated per unit volume
H	depth of a beam, lubricating film, channel or solution domain in general
h	heat transfer coefficient
h	distance between nodal points
h_r	distance between nodal points in radial direction
h_x, h_y	distance between nodal points in cartesian co-ordinate directions
I	second moment of area for bending
I	integral defined in equations 3.46
i	nodal point number
i_r	counter for circular rings of nodes and elements
i_x, i_y	counters for nodes and elements along rows in cartesian co-ordinate directions
i_θ	counter for nodes and elements around a circular ring
j	nodal point counter
K	ratio of concentric cylinder radii
K	overall stiffness matrix
K_{pq}	coefficient of K
$\underset{\sim}{K}_{pq}$	submatrix of K
$\widetilde{K}$	rectangular form of K
k	thermal conductivity
k	nodal point counter
k_m	element stiffness matrix
k_{rs}	coefficient of k
$\underset{\sim}{k}_{rs}$	submatrix of k
L	length of a beam
L	length of side of an equilateral triangle
L	vector storing the number of nonzero coefficients in the rows of matrix M
L_i	coefficient of L
L_m	length of an element
L_1, L_2	lengths of sides of elements on solution domain boundary
l_i, l_j, l_k	lengths of the sides of a triangular element
M	matrix storing original column numbers of coefficients of $\widetilde{K}$
M_i, M_j	moments applied internally to beam element at its nodes
M_{ij}	coefficient of M

m	element counter
N	bending moment
n	number of nodal points in a mesh
n	outward normal to the boundary of a solution domain
n_c	number of elements at centre of a circular mesh
n_q	number of a square in a mesh of right-angled triangles
n_r	number of nodes along a horizontal radius of a circular mesh
n_s	number of nodes per side of a triangular mesh
n_x, n_y	numbers of nodes per row of a rectangular mesh in the cartesian co-ordinate directions
P	an externally applied force
P_x, P_y, P_z	pressure gradients in cartesian co-ordinate directions
p	pressure
p	number of nonzero coefficients per row of overall stiffness matrix
Q	shear force
Q	volumetric flow rate
Q	externally applied force required to maintain a boundary restraint
q	number of iterations for convergence of the Gauss–Seidel solution process
R_m	vector of internal forces (and moments) applied to an element at its nodes
R_i	subvector of internally applied forces (and moments) at node i of an element
r	radial co-ordinate
r_1, r_2	relative efficiency parameters for methods of solving linear algebraic equations
$\bar{r}_m$	radius of the centroid of element m
S	mesh scale factor
S_x, S_y	summations involved in the Gauss–Seidel method, defined in equation 6.43
s	distance along a solution domain boundary
T	temperature
t	time
U_i, U_j, U_k	force components in x-direction applied internally to an element at its nodes
u	displacement or velocity in x-direction
$\bar{u}$	a mean value of u
V_i, V_j, V_k	force components in y-direction applied internally to an element at its nodes
V_x, V_z	velocity components of a boundary in cartesian co-ordinate directions
v	displacement or velocity in y-direction
$\bar{v}$	a mean value of v
W	width of a channel or solution domain in general

W	load applied to end of a cantilevered beam
W_i, W_j, W_k	force components in z-direction applied internally to an element at its nodes
w	displacement or velocity in z-direction
X, Y	global cartesian co-ordinates
$\bar{X}, \bar{Y}, \bar{Z}$	components of body forces per unit volume in cartesian co-ordinate directions
x, y, z	cartesian co-ordinates
α	coefficient of thermal expansion
α	prescribed value of dependent variable at a boundary
β	prescribed value of derivative normal to a boundary
γ	an angle
Δ	difference operator defining change in the subsequent quantity
Δ_m	element area
δ	overall vector of unknowns such as displacements or velocities
δ_i	unknown such as displacement or velocity at node i
$\boldsymbol{\delta_i}$	subvector of unknowns such as displacements or velocities at node i
$\bar{\boldsymbol{\delta}}_m$	vector of element unknowns such as displacements or velocities
ϵ	constant of proportionality in truncation error term
η	an unknown in a finite element analysis
θ	angle of rotation per unit length of a bar in torsion
θ	angular co-ordinate
$\boldsymbol{\theta}$	overall thermal force vector
$\boldsymbol{\theta}_m$	element thermal force vector
$\boldsymbol{\theta}_i$	subvector of thermal force components at node i of an element
θ_i, θ_j	rotations of ends of a beam element
$\theta_i, \theta_j, \theta_k$	angles at the corners of a triangular element
κ	permeability of a porous medium
λ	parameter defined in equation 2.86 or equation 3.41
μ	viscosity
ν	Poisson's ratio
π_P	dimensionless pressure gradient
π_Q	dimensionless flow rate
ρ	density
σ	stress
$\boldsymbol{\sigma}$	element stress vector
ϕ	angular co-ordinate
ϕ	sum of velocity components u and v
ϕ_1, ϕ_2	functions of position used in general harmonic and biharmonic equations 2.87 and 2.88
χ	stress function
χ	functional used in variational formulation of finite element analyses
ψ	stream function, or dependent variable in general

ω	vorticity
ω	over-relaxation factor
∇^2	harmonic operator
∇^4	biharmonic operator

Subscripts

A, B, C	points near a solution domain boundary
a to *h*	particular nodal points or elements
E, N, O, S, W	particular nodal points
i, j, k	nodal point numbers
l	counter used in Gauss–Seidel solution process
m	element number
p, q	node numbers involved in typical overall stiffness coefficient
p to *t*	particular nodal points on solution domain boundary
r	radial direction in polar co-ordinates
r, s	node numbers involved in typical element stiffness coefficient
T	thermal (strain) or truncation (error)
x, y, z	cartesian co-ordinate directions
θ	angular direction in polar co-ordinates
1, 2, 3	particular row or column numbers of element stiffness matrix
1, 2	referring to inner and outer of two concentric cylinders

Superscripts

(m)	due to element *m* (similarly *(e)* and *(f)*)
T	matrix transposition
*	modified quantity

Some Program Variable Names

The FORTRAN computer program variable names commonly used in the main text are defined in alphabetical order in the following list.

AI,AJ,AK	element dimensions a_i, a_j and a_k
ALPHA	coefficients of thermal expansion of the materials
AREA	areas of the elements
B	coefficients of the element dimension matrix
BI,BJ,BK	element dimensions b_i, b_j and b_k
BLANK	variable storing alphanumeric blank characters
BTD	coefficients of the matrix product $B^T D$
CASE	variable storing type of plane elastic problem
D	coefficients of element elastic property matrix
DELD	changes in the unknowns between successive cycles of Gauss–Seidel iteration
DELTA	unknowns such as displacements or velocities
DELTAT	temperature changes of the elements
E	Young's moduli
ERROR	relative error in Gauss–Seidel solution process
ESTIFF	coefficients of element stiffness matrix
ET	coefficients of element thermal strain vector
EXX,EXY,EYY	coefficients of element strain vector
F	coefficients of the vector of overall externally applied forces
FX,FY	coefficients of the vector of overall externally applied forces, in component form for x- and y-directions respectively
FXM,FYM	components of forces at nodes of an element due to distributed external force applied to one side
FXMOD,FYMOD	components of overall externally applied forces modified for body force and thermal effects

GXM,GYM	body force components acting at each node of an element
I	nodal point counter
IC	column counter for rectangular form of overall stiffness matrix
ICE	column counter for element stiffness matrix
ICOL	column counter for full (square) overall stiffness matrix
IFREQ	output frequency parameter in Gauss–Seidel solution process
IJK	numbers of the nodes of an element
IR	counter for circular rings of nodes and elements
IRE	row counter for element stiffness matrix
IROW	row counter for overall stiffness matrix
ITER	iteration counter in Gauss–Seidel solution process
ITH	counter for nodes and elements around a circular ring
IX,IY	counters for nodes and elements along rows in cartesian co-ordinate directions
J,K	nodal point counters
M	element counter
MATM	material numbers of the elements
MOUT	mesh data output control parameter
NAP	total numbers of nodal points adjacent to the nodes
NBC1P	number of nodes at which external forces are prescribed
NBC2F	number of distributed forces applied to solution domain
NBC3P	number of nodes at which displacement conditions are prescribed
NCEL	number of elements at centre of a circular mesh
NCOND	boundary condition type numbers for nodes at which displacement conditions are prescribed
NCYCLE	maximum number of iterations in Gauss–Seidel solution process
NEL	number of elements in the mesh
NEQN	number of equations to be solved
NMAT	number of different materials
NNP	number of nodal points in the mesh
NPA	numbers assigned to the nodal points adjacent to the nodes
NPB	numbers assigned to nodal points located on solution domain boundary

NPEQN	number of pairs of equations to be solved
NPI,NPJ,NPK	numbers assigned to the nodes of the elements
NREL	number of rings of elements in a circular mesh
NRPT	number of nodes along a horizontal radius of a circular mesh
NSPT	number of nodes per side of a triangular mesh
NTHPT	number of nodes or elements around a circular ring
NU	Poisson's ratios of the materials
NXEL,NYEL	numbers of elements per row of a rectangular mesh in the cartesian co-ordinate directions
NXPT,NYPT	numbers of nodal points per row of a rectangular mesh in the cartesian co-ordinate directions
OKXX,OKXY,OKYX,OKYY	coefficients of overall stiffness submatrices
ORELAX	over-relaxation factor
OSTIFF	coefficients of overall stiffness matrix
PHI1	mean values of function ϕ_1 for the elements
PI	π
PX,PY	components of distributed external forces per unit surface area
RHO	densities of the materials
SFXX,SFXY,SFYX,SFYY	coefficients of self-flexibility submatrices
SIDE	length of side of an element subjected to a distributed force
SIGXX,SIGXY,SIGYY	coefficients of element stress vector
SUMD	summed magnitudes of the unknowns
SUMDD	summed magnitudes of the changes in the unknowns between successive cycles of iteration
TANG	tangents of angles of directions of free motion for nodes at which displacement conditions are prescribed
THETA	angular co-ordinate θ
THETAM	coefficients of element thermal force vector
TITLE	alphanumeric title for the problem
TOLER	convergence tolerance for Gauss–Seidel solution process
U,V	displacement components in cartesian co-ordinate directions
UPRES,VPRES	prescribed values of displacement components
X,Y	global cartesian co-ordinates of the nodal points

NEQN	number of pairs of equations to be solved
NPI,NPJ,NPK	numbers assigned to the nodes of the elements
NPHT	number of rings of elements in a circular mesh
NPPT	number of nodes along a horizontal radius of a circular mesh
PSPT	number of nodes per side of a triangular mesh
NTHET	number of rings of elements around a circular ring
NU	Poisson's ratio of the materials
NXRT,NYRT	numbers of elements per row of a rectangular mesh in the cartesian coordinate directions
NXPT,NYPT	numbers of additional rows of elements in the cartesian or evaluate directions
OKA,OLXY,OK,Y,OKX,Y	coefficients of overall stiffness submatrices
OELI,AK	over-relaxation factor
OSTIF	coefficient of overall stiffness matrix
PHI	mean values of function ϕ for the elements
PI	π
PX,PY	components of distributed external forces per unit surface area
RHO	densities of the materials
SRXX,SDXX,PI,YX,SLYY	coefficients of stiffness-like submatrices
SIDE	length etc. of an element subjected to a distributed force
SLX,Y,SLKY,SLKY	coefficients of element stiffness submatrices
SUMF	summed magnitudes of the unknowns
SUMD	summed magnitude of the changes in the unknowns between successive cycles of iteration
TAVG	tangent of angles of direction of the motion for nodes at which displacement conditions are prescribed
THETA	angular coordinate θ
THETAM	coefficients of element thermal force vector
TITLE	alphanumeric title for the problem
TOLER	convergence tolerance for Gauss-Seidel solution process
U,V	displacement components in certain coordinate directions
UPRES,VPRES	prescribed values of displacement components
X,Y	global cartesian coordinates of the nodal points

1 Introduction and Structural Analysis

Many practical problems in engineering are either extremely difficult or impossible to solve by conventional analytical methods. Such methods involve finding mathematical equations which define the required variables. For example, the distribution of stresses and displacements in a solid component, or of pressure and velocities in the flow of a fluid, might be required. In the past it was common practice to simplify such problems to the point where an analytical solution could be obtained which, it was hoped, bore some resemblance to the solution of the real problem. Because of the uncertainties involved in such a procedure, large 'factors of safety' were introduced, which could often be described more accurately as factors of ignorance. With the relatively recent advent of high-speed electronic digital computers, however, the emphasis in engineering analysis has moved towards more versatile numerical methods. One class of such methods has been given the name finite element methods.

Finite element methods originated in the field of structural analysis and were widely developed and exploited in the aerospace industries during the '50s and '60s. Such methods are firmly established in civil and aeronautical engineering, as witnessed by the authorship of some of the books on the subject, such as Zienkiewicz (1971), Desai and Abel (1972) and Nath (1974). Finite element methods are also widely used by mechanical engineers, particularly for the analysis of stress in solid components. Their success has been such that experimental methods involving brittle coatings, strain gauges or photoelastic effects are now to some extent obsolete. Problems in fluid mechanics and heat transfer are, however, much less commonly solved by finite element methods. One possible reason for this is that such problems are made difficult not so much by geometric complexities as by the nature of the physical processes involved. For example, relatively little attention has so far been given to the finite element solution of problems where thermal convection is important.

All finite element methods involve dividing the physical systems, such as structures, solid or fluid continua, into small subregions or elements. Each element is an essentially simple unit, the behaviour of which can be readily analysed. The complexities of the overall systems are accommodated by using large numbers of elements, rather than by resorting to the sophisticated mathematics required by many analytical solutions. One of the main attractions of finite element methods is the ease with which they can be applied to problems involving geometrically complicated systems. The price that must be paid for flexibility and simplicity of individual elements is in the amount of numerical computation required. Very large sets of simultaneous algebraic equations have to be solved, and this can only be done economically with the aid of digital computers. Fortunately, finite element methods may be readily programmed for this purpose.

The object of this book is to provide an introduction to finite element methods, particularly those applicable to continuum mechanics problems of stress analysis, fluid mechanics and heat transfer. For the most part, only the simplest of such methods are described in detail. Problems considered are mainly of the equilibrium type, involving either statically loaded components or steady fluid flows. Also, although finite element methods are applicable to either one-, two- or three-dimensional problems, the main emphasis here is on two-dimensional ones. A number of practical case studies, including computer programs and numerical results, are examined in detail.

After a brief review of the subject of computer programming, the remainder of this chapter is devoted to simple types of structural analysis which serve to introduce finite element methods. In chapter 2 a wide range of continuum mechanics problems is reviewed, and finite element methods for solving them are described in subsequent chapters.

1.1 Computer Programming

In this book a considerable amount of attention is given to computer programs for solving engineering problems using finite element methods. The programming language used is FORTRAN, whose name is derived from FORmula TRANslation, and which is particularly suitable for engineering and scientific applications. While various levels of FORTRAN have been developed, the version used here is FORTRAN IV as described by, for example, McCracken (1972). The application of FORTRAN programming to engineering problems in general, and to elementary finite element methods in particular, is described by Fenner (1974).

Although FORTRAN is largely independent of the particular computer used, there are a few features which are machine dependent. For example, input READ statements and output WRITE statements refer to particular magnetic tape numbers. All the examples in this book use the numbers 5 and 6 for input and output respectively. The computer user normally submits his program and data in the form of a deck of punched cards, or occasionally punched paper tape, the information from which is read onto the magnetic input tape. The results stored on the magnetic output tape are normally recorded on paper by a line printer. The fast core store in a computer is composed of a large number of 'words' or storage registers, each of which may contain, for example, a number being used in the calculation. Most modern computers work in binary arithmetic and individual words contain a particular number of binary digits or 'bits'. This number varies considerably from one computer to another, but is usually between 24 and 60. If the number of bits is small the precision of stored numbers is relatively low and significant roundoff errors may be accumulated in the course of a calculation. All the case studies described in this book were run on a computer with a sixty-bit word length, giving at least twelve-decimal digit precision, and allowing the storage of up to ten alphanumeric characters per word.

The style of writing FORTRAN programs should be such as to make the coding simple to follow and check, and at the same time efficient in terms of execution time and core storage. With these requirements in mind the programs in this book use variable names which are readily identifiable with the physical or mathematical quantities they represent. Whenever possible the same names are used throughout, their definitions being listed at the beginning of the book. Large programs are divided into shorter subprograms which can be written, developed and tested separately. Also, comment statements are used liberally, both to explain the coding and to separate successive sets of statements for improved readability. For the same reason, a uniform system of statement numbering is used within each subprogram. Those executable statements requiring numbers are numbered in sequence from 1, while input and output FORMAT statements are numbered from 51 and 61 respectively.

1.2 Structural Analysis

The analysis of engineering structures provides a natural introduction to finite element methods. In this section the analysis of very simple structures is considered, and a case study concerning a simple rigid-jointed structure is described in section 1.3.

1.2.1 Pin-jointed structures Among the simplest types of structures are frameworks consisting of relatively long thin members pin-jointed at their ends. The members may be subjected to tension or compression, but not to bending or torsion. They form natural finite elements of the overall structure. Fenner (1974) describes a method and computer program for analysing plane pin-jointed frameworks which are either statically determinate or statically indeterminate. It is therefore appropriate to turn to rigid-jointed structures to provide a deliberately simple introductory example for the purposes of this book.

1.2.2 Rigid-jointed structures Many structures involve members which are rigidly jointed together, and which may be subjected to bending and torsion in addition to tension or compression. In many cases the strains and displacements due to bending are very much larger than those due to tensile loads. Attention can therefore be confined to bending effects, and the members treated as being perfectly rigid in tension or compression.

As an example of a one-dimensional rigid-jointed structure, consider the cantilevered beam shown in figure 1.3, which is the subject of the case study presented in section 1.3. Provided the length of the beam is large compared with its depth, it can be divided into a number of simple uniform beam elements of the type shown in figure 1.1. The extent to which a series of such parallel elements can be used to approximate the behaviour of a tapered beam is examined in section 1.3.

The first stage in the analysis is to examine the behaviour of individual elements. Local co-ordinates x and y are chosen to be respectively along and normal to the neutral axis of the typical element shown in figure 1.1. All the elements are numbered, and the number of this element is m. Its length, second moment of area for bending in the $x-y$ plane and Young's modulus are L_m, I_m and E_m respectively. The element is joined to its immediate neighbours via 'nodes' at its ends which are numbered i and j. When the structure is loaded, the displacement in the y-direction and clockwise rotation of the typical point i are v_i and θ_i, and the corresponding force component and moment applied to the element are V_i and M_i.

One way to proceed with the analysis is to assume a suitable form of variation for the displacement in the y-direction along the element

$$v(x) = C_1 + C_2 x + C_3 x^2 + C_4 x^3 \tag{1.1}$$

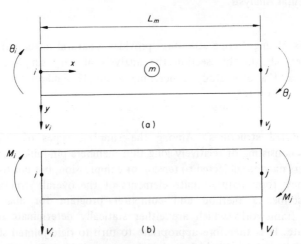

Figure 1.1   A simple beam element: (a) displacements and rotations; (b) forces and moments

where C_1 to C_4 are constant for the particular element. It is convenient to choose a polynomial function, and a cubic form allows the following modes of motion and deformation of the element.

(1) Rigid body translation ($C_1 \neq 0$)
(2) Rigid body rotation ($C_2 \neq 0$)
(3) Bending strain ($C_3 \neq 0$)
(4) Shear strain (due to shear forces in the y-direction: $C_4 \neq 0$)

The use of a function involving four parameters allows their values to be found in terms of the four nodal point displacements v_i, θ_i, v_j and θ_j. The clockwise rotation at any position along the element is given by

$$\theta = \frac{dv}{dx} = C_2 + 2C_3x + 3C_4x^2 \tag{1.2}$$

Hence

$$v_i = v(0) = C_1 \tag{1.3}$$

$$\theta_i = \theta(0) = C_2 \tag{1.4}$$

$$v_j = v(L_m) = C_1 + C_2L_m + C_3L_m{}^2 + C_4L_m{}^3 \tag{1.5}$$

$$\theta_j = \theta(L_m) = C_2 + 2C_3L_m + 3C_4L_m{}^2 \tag{1.6}$$

The values of C_1 and C_2 are given directly by equations 1.3 and 1.4, those of C_3 and C_4 being obtained from equations 1.5 and 1.6 as

$$C_3 = -\frac{3v_i}{L_m{}^2} - \frac{2\theta_i}{L_m} + \frac{3v_j}{L_m{}^2} - \frac{\theta_j}{L_m} \tag{1.7}$$

$$C_4 = \frac{2v_i}{L_m{}^3} + \frac{\theta_i}{L_m{}^2} - \frac{2v_j}{L_m{}^3} + \frac{\theta_j}{L_m{}^2} \tag{1.8}$$

From the simple theory of bending, the hogging bending moment N and shear force Q defined as in figure 1.2 are given by

$$N = E_mI_m \frac{d^2v}{dx^2} = E_mI_m(2C_3 + 6C_4x) \tag{1.9}$$

$$Q = \frac{dN}{dx} = 6E_mI_mC_4 \tag{1.10}$$

The forces and moments applied to the element at its nodes may therefore be

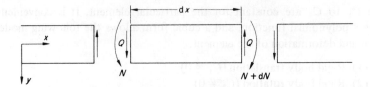

Figure 1.2   Bending moment and shear force in the simple bending of a beam

expressed in matrix form as

$$
\begin{bmatrix} V_i \\ M_i \\ V_j \\ M_j \end{bmatrix} = \begin{bmatrix} Q(0) \\ -N(0) \\ -Q(L_m) \\ N(L_m) \end{bmatrix} = E_m I_m \begin{bmatrix} 0 & 6 \\ -2 & 0 \\ 0 & -6 \\ 2 & 6L_m \end{bmatrix} \begin{bmatrix} C_3 \\ C_4 \end{bmatrix}
\tag{1.11}
$$

and using equations 1.7 and 1.8 for C_3 and C_4

$$
\begin{bmatrix} V_i \\ M_i \\ V_j \\ M_j \end{bmatrix} = \frac{E_m I_m}{L_m{}^3} \begin{bmatrix} 12 & 6L_m & -12 & 6L_m \\ 6L_m & 4L_m{}^2 & -6L_m & 2L_m{}^2 \\ -12 & -6L_m & 12 & -6L_m \\ 6L_m & 2L_m{}^2 & -6L_m & 4L_m{}^2 \end{bmatrix} \begin{bmatrix} v_i \\ \theta_i \\ v_j \\ \theta_j \end{bmatrix}
\tag{1.12}
$$

that is

$$
R_m = k_m \delta_m
\tag{1.13}
$$

where R_m and δ_m are the element force and displacement vectors and k_m is the element stiffness matrix, which is symmetric.

The next stage in the analysis is to add together the force–displacement characteristics of the individual elements to determine the behaviour of the overall structure. For general rigid-jointed structures in which the members and elements are at arbitrary angles to one another it would be necessary to transform from the local co-ordinates used above for the individual elements to the 'global' co-ordinates for the overall structure (see, for example, Desai and Abel (1972)). For problems of the present straight beam type, however, it is possible to proceed immediately with the assembly process. Since the elements are connected at the nodes, the displacements (both linear and rotational) of a particular node are the same for every element connected to it. Also, the conditions for equilibrium of the structure may be expressed in words and symbols as

$$\begin{pmatrix} \text{externally applied forces} \\ \text{and moments at the nodes} \end{pmatrix} = \Sigma \begin{pmatrix} \text{forces and moments on} \\ \text{the elements at these nodes} \end{pmatrix}$$

$$F = \Sigma R_m = \Sigma k_m \delta_m = K \delta \tag{1.14}$$

where K is the overall stiffness matrix, and the vectors F and δ contain the externally applied forces and moments, and the corresponding linear and rotational displacements. For example, if there are n nodes, δ contains $v_1, \theta_1,$ $v_2, \theta_2, \ldots, v_n, \theta_n$, in this order. The summation in equation 1.14 is for all the elements in the structure, and the coefficients of the overall stiffness matrix are assembled from those of the stiffness matrices of the individual elements. The matrix K is 'sparse' (having relatively few nonzero coefficients) because not more than two elements are connected to any one nodal point.

Let K_{pq} and k_{rs} be typical coefficients of the overall and element stiffness matrices respectively, where p and q lie in the range 1 to $2n$, while r and s lie in the range 1 to 4. The subscripts p and r are matrix row numbers, while q and s are column numbers. Now k_{rs} can be interpreted as the force or moment that must be applied to the typical element at the node and in the direction or sense corresponding to the rth coefficient of the element force vector to cause a unit linear or rotational displacement at the node and in the direction or sense corresponding to the sth coefficient of the element displacement vector. A similar interpretation can be applied to K_{pq} in terms of the pth coefficient of F and the qth coefficient of δ. The process of assembling the overall stiffness coefficients takes the form of

$$K_{pq} = \Sigma k_{rs} \tag{1.15}$$

where the row and column numbers are equivalent. For example, if $r < 3$ the nodal point number concerned is i, and the equivalent value of p is $2(i-1)+r$. Similarly, if $s \geqslant 3$ the nodal point number concerned is j, and the equivalent value of q is $2(j-1) + (s-2)$. Assembly is complete when the sixteen coefficients of each and every element stiffness matrix have been added to the relevant overall stiffness coefficients.

Before linear equations 1.14 can be solved for the displacements, the restraint conditions appropriate to the particular problem must be applied. The values of at least two displacements are normally prescribed. Such conditions can be applied by modifying the corresponding equations, for example, by the method described in section 1.3.3. Given the computed displacements, the internal forces and bending moments at the nodal points can be found with the aid of equations 1.12.

In some problems involving rigid-jointed structures the members are subjected to loads distributed along their lengths. For example, the weights of the members may cause significant displacements. Such loadings cannot be accommodated exactly by the present method which assumes the cubic displacement function given by equation 1.1. If the differential element shown in figure 1.2 is subject to a uniformly distributed vertical load of q per unit length, then

$$q = \frac{dQ}{dx} = E_m I_m \frac{d^4 v}{dx^4} \qquad (1.16)$$

which would be zero. Such a uniformly distributed load can be accommodated approximately, however, by taking the total load applied to an element, qL_m, and dividing it equally between its two nodes as point loads.

1.3 Case Study: Bending of a Tapered Beam

Figure 1.3 shows a cantilevered beam carrying a vertical load at its free end. The cross-section of the beam is rectangular with constant width B (normal to the $x-y$ plane shown) and depth H which tapers linearly from H_0 to $\frac{1}{2}H_0$.

1.3.1 Problem specification The vertical and rotational displacements of the free end of the beam are to be computed using the method described in section 1.2.2, and compared with the exact values obtained from the analytical solution to the problem. The length of the beam is 5 m, width 0.1 m, and the initial depth H_0 is 0.3 m. Young's modulus of the steel from which it is made is 208 GN m^{-2}, and the end load W is 10 kN. The beam is to be divided into ten elements of equal length.

1.3.2 Analytical solution Using the simple theory of bending, the hogging bending moment at a distance x from the fixed end of the beam is

$$N = EI \frac{d^2 v}{dx^2} = W(L - x) \qquad (1.17)$$

where E is Young's modulus and I is the local second moment of area. For the present rectangular cross-sectioned beam

$$I = \frac{BH^3}{12} \qquad (1.18)$$

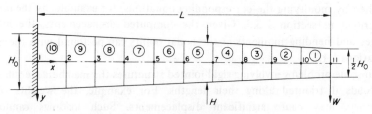

Figure 1.3   A cantilevered tapered beam

where

$$H = H_0 \left(1 - \frac{x}{2L}\right) \tag{1.19}$$

Therefore, if I_0 is the second moment of area at $x = 0$

$$v''(x) = \frac{W}{EI_0}(L - x)\left(1 - \frac{x}{2L}\right)^{-3}$$

$$= \frac{WL}{EI_0}\left[2\left(1 - \frac{x}{2L}\right)^{-2} - \left(1 - \frac{x}{2L}\right)^{-3}\right] \tag{1.20}$$

The slope and deflection may be found by integration as

$$v'(x) = \frac{WL^2}{EI_0}\left[4\left(1 - \frac{x}{2L}\right)^{-1} - \left(1 - \frac{x}{2L}\right)^{-2}\right] + A_1$$

$$v(x) = \frac{WL^3}{EI_0}\left[-8\ln\left(1 - \frac{x}{2L}\right) - 2\left(1 - \frac{x}{2L}\right)^{-1}\right] + A_1 x + A_2$$

where A_1 and A_2 are integration constants, which may be found with the aid of the boundary conditions

$$v(0) = v'(0) = 0 \tag{1.21}$$

to give

$$v'(x) = \frac{WL^2}{EI_0}\left[4\left(1 - \frac{x}{2L}\right)^{-1} - \left(1 - \frac{x}{2L}\right)^{-2} - 3\right] \tag{1.22}$$

$$v(x) = \frac{WL^3}{EI_0}\left[-8\ln\left(1 - \frac{x}{2L}\right) - 2\left(1 - \frac{x}{2L}\right)^{-1} - \frac{3x}{L} + 2\right] \tag{1.23}$$

With the present data, $I_0 = 2.25 \times 10^{-4}$ m^4, and the required displacements are

$$v'(L) = \frac{WL^2}{EI_0} = 0.5342 \times 10^{-2} \text{ rad} \tag{1.24}$$

$$v(L) = \frac{WL^3}{EI_0}(-8\ln 0.5 - 5) = 0.1456 \times 10^{-1} \text{ m} \tag{1.25}$$

1.3.3 Numerical solution   Figure 1.4 shows a FORTRAN main program for analysing the bending behaviour of straight beams of arbitrary cross-section. The method of formulation is that described in section 1.2.2, and the resulting set of simultaneous linear algebraic equations is solved by gaussian elimination in a subprogram named ELIMIN. Both the elimination method and subprogram are described in appendix A.

```
C   PROGRAM FOR STRUCTURAL ANALYSIS OF A STRAIGHT BEAM.
C
        DIMENSION  NREST(51),X(51),NPI(50),NPJ(50),E(50),SMA(50),
      1            ESTIFF(4,4),OSTIFF(102,103),DELTA(102),VTH(2,51),F(102)
        REAL  L(50),LOAD(2,51)
        EQUIVALENCE  (F(1),LOAD(1,1)),(DELTA(1),VTH(1,1))
C
C   INPUT AND TEST THE NUMBER OF ELEMENTS.
   1    READ(5,51) NEL
  51    FORMAT(I5)
        IF(NEL.GT.0.AND.NEL.LE.50) GO TO 2
        STOP
   2    WRITE(6,61)
  61    FORMAT(39H1STRUCTURAL ANALYSIS OF A STRAIGHT BEAM)
C
C   INPUT THE NODAL POINT DATA.
        NNP=NEL+1
        READ(5,52) (I,NREST(I),X(I),LOAD(1,I),LOAD(2,I),N=1,NNP)
  52    FORMAT(2I5,F10.0,2E15.5)
C
C   INPUT THE ELEMENT DATA.
        READ(5,53) (M,NPI(M),NPJ(M),E(M),SMA(M),N=1,NEL)
  53    FORMAT(3I5,2E15.5)
C
C   PREPARE TO SUM THE STIFFNESS COEFFICIENTS.
        NEQN=2*NNP
        DO 3 IROW=1,NEQN
        DO 3 ICOL=1,NEQN
   3    OSTIFF(IROW,ICOL)=0.
        DO 6 M=1,NEL
C
C   FORM THE STIFFNESS MATRIX FOR EACH ELEMENT.
        I=NPI(M)
        J=NPJ(M)
        L(M)=X(J)-X(I)
        FACT=E(M)*SMA(M)/L(M)**3
        ESTIFF(1,1)=FACT*12.
        ESTIFF(1,2)=FACT*6.*L(M)
        ESTIFF(1,3)=-ESTIFF(1,1)
        ESTIFF(1,4)= ESTIFF(1,2)
        ESTIFF(2,1)= ESTIFF(1,2)
        ESTIFF(2,2)=FACT*4.*L(M)**2
        ESTIFF(2,3)=-ESTIFF(2,1)
        ESTIFF(2,4)=FACT*2.*L(M)**2
        DO 4 ICE=1,4
   4    ESTIFF(3,ICE)=-ESTIFF(1,ICE)
        DO 5 ICE=1,3
   5    ESTIFF(4,ICE)=ESTIFF(ICE,4)
        ESTIFF(4,4)=ESTIFF(2,2)
C
C   ADD ELEMENT STIFFNESS TO OVERALL STIFFNESS.
        DO 6 IRE=1,4
        DO 6 ICE=1,4
        IF(IRE.LT.3) IROW=2*(I-1)+IRE
        IF(IRE.GE.3) IROW=2*(J-1)+IRE-2
        IF(ICE.LT.3) ICOL=2*(I-1)+ICE
        IF(ICE.GE.3) ICOL=2*(J-1)+ICE-2
   6    OSTIFF(IROW,ICOL)=OSTIFF(IROW,ICOL)+ESTIFF(IRE,ICE)
C
C   APPLY THE RESTRAINTS.
        DO 9 I=1,NNP
        IF(NREST(I).EQ.0) GO TO 9
        DO 8 N=1,2
        IF(NREST(I).NE.N.AND.NREST(I).NE.3) GO TO 8
        IROW=2*(I-1)+N
        DO 7 ICOL=1,NEQN
   7    IF(IROW.NE.ICOL) OSTIFF(IROW,ICOL)=0.
        LOAD(N,I)=0.
   8    CONTINUE
   9    CONTINUE
```

Figure 1.4   Main program for the analysis of beams in bending

```
C
C   EXTEND THE OVERALL STIFFNESS MATRIX TO INCLUDE THE FORCE VECTOR.
        DO 10 IROW=1,NEQN
   10   OSTIFF(IROW,NEQN+1)=F(IROW)
C
C   SOLVE THE LINEAR EQUATIONS.
        CALL  ELIMIN(OSTIFF,DELTA,NEQN,102,103,DET,RATIO)
C
C   OUTPUT THE RESULTS.
        WRITE(6,62) (M,NPI(M),NPJ(M),E(M),SMA(M),L(M),M=1,NEL)
   62   FORMAT(52H0     M     I     J    MODULUS  2ND MOM AREA    LENGTH    /
       1      (1X,3I5,3E12.4))
        WRITE(6,63) (I,NREST(I),X(I),(LOAD(N,I),VTH(N,I),N=1,2),I=1,NNP)
   63   FORMAT(71H0    I  REST     X          LOAD        DEFLN        MOMENT
       1       ROTATION   / (1X,2I5,5E12.4))
        GO TO 1
        END
```

Figure 1.4   Continued

The DIMENSION statement in figure 1.4 allows for beams involving up to 50 elements and 51 nodal points to be analysed. The arrays F, OSTIFF and DELTA are used to store the externally applied nodal point forces and moments, the coefficients of the overall stiffness matrix, and the linear and rotational displacements of the nodal points respectively. Note that OSTIFF is allowed an extra column so that it can be extended to include the nodal point forces and moments, as required by ELIMIN. While F and DELTA are in single subscript form, for some purposes double subscripts are more convenient: one for the nodal point number and the other to indicate translation or rotation. Hence, arrays LOAD and VTH are introduced and refer to the same storage registers as F and DELTA by virtue of the EQUIVALENCE statement. The variables VTH(1,I) and VTH(2,I), for example, store the values of v_i and θ_i respectively, where i is the value of the counter I. In FORTRAN the components of an array with more than one subscript are stored in an order such that the first subscript varies most rapidly.

The array X is used to store the co-ordinates of the nodal points, while NREST stores integer numbers which define the type of restraint condition applied to each point. With the program in its present form, a zero value of NREST(I) means that the point whose number is given by the value of I is unrestrained, while values of one, two or three mean that either the linear, rotational or both displacements of the point are prescribed as zero. The program could be readily modified to allow nonzero values to be prescribed. The arrays NPI and NPJ are used to store the numbers of the nodal points (i and j for the typical element shown in figure 1.1) at the ends of the elements, while SMA, E and L store the element second moments of area, Young's moduli and lengths. Finally, the array ESTIFF is used to store the coefficients of the element stiffness matrices.

Other variables used in the main program include NEL, NNP and NEQN for the number of elements, nodal points and equations respectively, while the counters I and J are used for nodal point numbers, and M for element numbers. IROW and ICOL are used for row and column numbers in the overall stiffness matrix, while IRE and ICE serve the same purpose for the element stiffness matrices (p, q, r and s are the equivalent counters in equation 1.15).

After the number of elements has been read in, the number, restraint condition, co-ordinate, and externally applied force and moment for each node are read in, from a new data card for each node. Then the number, the associated nodal point numbers, Young's modulus and second moment of area for each element are read in. For the present problem, the second moments of area are calculated as described in section 1.3.4. The coefficients of the overall stiffness matrix are first set to zero in preparation for the assembly process. Then, for each element in turn, the numbers of the associated nodal points are assigned to I and J, and the coefficients of the element stiffness matrix are computed according to equation 1.12. Each of these coefficients is added to the appropriate coefficient of the overall stiffness matrix, as indicated by equation 1.15.

The restraint conditions are applied according to the numbers stored in NREST. If a nonzero value is detected for a particular point numbered i, either the linear, rotational or both displacements of that point are fixed according to whether the value is one, two or three, by altering the corresponding equations to give $v_i = 0$ or $\theta_i = 0$. This is done by setting all the coefficients in the relevant row of the overall stiffness matrix not on the diagonal to zero, together with the corresponding externally applied force or moment.

In order that use can be made of subprogram ELIMIN, the overall stiffness matrix is extended to include the externally applied forces and moments. The subprogram arguments DET and RATIO, which are not used by the present main program, are discussed in appendix A. Both the input data and the computed displacements are written out before the program returns to read the data for a new problem. Termination of execution occurs when an unacceptable number of elements is encountered, and may therefore be achieved by the use of a blank card after the data for a particular problem.

There are a number of additions and improvements that could be made to the present program. As already mentioned, more general restraint conditions could be applied. The data required for the elements and nodal points could be generated within the program in order to reduce the amount of information which must be punched on data cards. Tests for the validity of the data read in can be devised. For example, a test for negative computed element lengths would provide a simple check on numbering procedures. Such refinements are omitted in order to keep this introductory example as simple as possible.

It should be noted that the program uses large amounts of computer core storage, over ten thousand words being required to store the overall stiffness matrix for a problem involving no more than fifty elements. This is because the gaussian elimination method in its basic form requires the entire stiffness matrices to be stored. In the present type of problem these matrices are very sparse, with not more than six nonzero coefficients per row or column, usually arranged either on or adjacent to the diagonals of the matrices. More

STRUCTURAL ANALYSIS OF A STRAIGHT BEAM

M	I	J	MODULUS	2ND MOM AREA	LENGTH
1	10	11	0.2080E+12	0.3256E-04	0.5000E+00
2	9	10	0.2080E+12	0.4277E-04	0.5000E+00
3	8	9	0.2080E+12	0.5493E-04	0.5000E+00
4	7	8	0.2080E+12	0.6920E-04	0.5000E+00
5	6	7	0.2080E+12	0.8574E-04	0.5000E+00
6	5	6	0.2080E+12	0.1047E-03	0.5000E+00
7	4	5	0.2080E+12	0.1263E-03	0.5000E+00
8	3	4	0.2080E+12	0.1507E-03	0.5000E+00
9	2	3	0.2080E+12	0.1781E-03	0.5000E+00
10	1	2	0.2080E+12	0.2085E-03	0.5000E+00

I	REST	X	LOAD	DEFLN	MOMENT	ROTATION
1	3	0.E+00	0.E+00	0.E+00	0.E+00	0.E+00
2	0	0.5000E+00	0.E+00	0.1393E-03	-0.E+00	0.5475E-03
3	0	0.1000E+01	0.F+00	0.5593E-03	-0.E+00	0.1121E-02
4	0	0.1500E+01	0.E+00	0.1273E-02	-0.E+00	0.1719E-02
5	0	0.2000E+01	0.E+00	0.2291E-02	-0.E+00	0.2338E-02
6	0	0.2500E+01	0.E+00	0.3622E-02	-0.E+00	0.2969E-02
7	0	0.3000E+01	0.F+00	0.5270E-02	-0.E+00	0.3600E-02
8	0	0.3500E+01	0.E+00	0.7229E-02	-0.E+00	0.4208E-02
9	0	0.4000E+01	0.E+00	0.9479E-02	-0.E+00	0.4755E-02
10	0	0.4500E+01	0.F+00	0.1197E-01	-0.E+00	0.5176E-02
11	0	0.5000E+01	0.1000E+05	0.1462E-01	-0.E+00	0.5361E-02

Figure 1.5   Results from beam bending analysis program

sophisticated and economical versions of the elimination process are available for such banded matrices. General techniques for solving the linear algebraic equations arising in finite element methods are discussed in later chapters.

1.3.4 Results   Figure 1.5 shows the results obtained with the specified data, and the element arrangement shown in figure 1.3. Although the nodes and elements are numbered in sequence, the orders could be arbitrary. The printed results show the nodal point numbers, modulus, second moment of area and length for each of the ten elements. The second moment of area is taken as the value at the centre of the element. For example, for element number 1, whose centre is at x = 4.75 m, equations 1.18 and 1.19 give

$$I = I_0 \left(1 - \frac{x}{2L} \right)^3 = 0.3256 \times 10^{-4} \ \text{m}^4$$

Then for each of the eleven nodal points, the restraint condition number, co-ordinate, applied force, deflection, applied moment and rotation are printed.

Compared with the exact values given by equations 1.24 and 1.25, the computed end deflection and rotation (for node number 11) are both about 0.4 per cent in error. Such accuracy is adequate for most practical purposes, but could be improved by increasing the number of elements or by using better average values for the second moments of area. In the present problem all the elements are of the same length, whereas the program is equally capable of accommodating elements of varying length, a facility which can be of considerable value in more general problems. Also, problems can be handled where the modulus of the material varies with position along the beam.

The main purpose of this case study is to serve as an introduction to the formulation and programming of finite element methods. Clearly, the present approach can be extended to more complex structural elements, subjected to forces and moments in all three co-ordinate directions. In the remainder of this book, however, attention is concentrated on problems in continuum mechanics.

2 Continuum Mechanics Problems

The concept of treating solids and fluids as though they are continuous media, rather than composed of discrete molecules, is one that is widely used in most branches of engineering. The purpose of this chapter is to review the principles of continuum mechanics and to apply them to a number of typical physical problems. It is shown that the resulting mathematical equations are of very similar types, and can therefore be solved by similar methods. Finite difference methods are reviewed, while later chapters in this book are devoted to finite element methods for solving such continuum problems.

2.1 Continuum Mechanics Equations

The mathematical equations describing the deformation of solids are very similar in form to those for the flow of fluids. The principal difference is that, while solid behaviour is formulated in terms of displacements and strains, the equivalent variables for fluids are velocities and strain rates. Another distinction is that the co-ordinate systems used are usually fixed relative to the material in the case of solids, but are fixed in space in the case of fluids. This means that the equations are only applicable to small strains in solid bodies such that their overall geometries are not significantly affected. No such restriction applies to strain rates in fluids.

The equations are displayed in their full three-dimensional cartesian forms. Other co-ordinate systems are rarely used in this book and most of the problems considered are essentially two-dimensional, permitting further simplification of the equations. Full derivations are given by, for example, Ford (1963) or Bird *et al.* (1960).

2.1.1 Stresses and strains   Throughout this book the symbol σ is used to denote stress. Individual components of stress are indicated by double subscripts as follows

direct stresses: $\sigma_{xx}, \sigma_{yy}, \sigma_{zz}$

shear stresses: $\sigma_{xy}, \sigma_{yz}, \sigma_{zx}, \sigma_{yx}, \sigma_{zy}, \sigma_{xz}$

The first subscript defines the direction of the stress and the second one denotes the direction of the outward normal to the surface on which it acts, as shown in figure 2.1. According to this convention tensile stresses are positive. For rotational equilibrium to be maintained the shear stresses must be complementary

$$\sigma_{xy} = \sigma_{yx}, \quad \sigma_{yz} = \sigma_{zy}, \quad \sigma_{zx} = \sigma_{xz} \tag{2.1}$$

The components of displacement (for solids) or velocity (for fluids) in the co-ordinate directions x, y and z are denoted by u, v and w respectively. Using the same double subscript notation, the direct and shear components of strain or strain rate may be defined as

$$e_{xx} = \frac{\partial u}{\partial x}, \quad e_{yy} = \frac{\partial v}{\partial y}, \quad e_{zz} = \frac{\partial w}{\partial z} \tag{2.2}$$

$$e_{xy} = e_{yx} = \frac{\partial u}{\partial y} + \frac{\partial v}{\partial x} \tag{2.3}$$

$$e_{yz} = e_{zy} = \frac{\partial v}{\partial z} + \frac{\partial w}{\partial y} \tag{2.4}$$

$$e_{zx} = e_{xz} = \frac{\partial w}{\partial x} + \frac{\partial u}{\partial z} \tag{2.5}$$

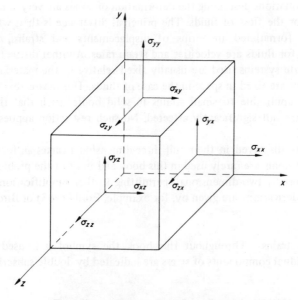

Figure 2.1   Cartesian stress components

Alternative definitions for the shear components are sometimes used, particularly in fluid mechanics. These involve the introduction of a factor of ½: for example

$$e_{xy} = e_{yx} = \frac{1}{2}\left(\frac{\partial u}{\partial y} + \frac{\partial v}{\partial x}\right) \tag{2.6}$$

Throughout this book the definitions given in equations 2.2 to 2.5 are used in applications to both solids and fluids.

2.1.2 Equilibrium equations If inertia forces are negligibly small the differential equations of equilibrium for the three co-ordinate directions can be expressed as

$$\frac{\partial \sigma_{xx}}{\partial x} + \frac{\partial \sigma_{xy}}{\partial y} + \frac{\partial \sigma_{xz}}{\partial z} + \bar{X} = 0 \tag{2.7}$$

$$\frac{\partial \sigma_{yx}}{\partial x} + \frac{\partial \sigma_{yy}}{\partial y} + \frac{\partial \sigma_{yz}}{\partial z} + \bar{Y} = 0 \tag{2.8}$$

$$\frac{\partial \sigma_{zx}}{\partial x} + \frac{\partial \sigma_{zy}}{\partial y} + \frac{\partial \sigma_{zz}}{\partial z} + \bar{Z} = 0 \tag{2.9}$$

where $\bar{X}$, $\bar{Y}$ and $\bar{Z}$ are the local components of the body forces per unit volume acting on the continuum in the co-ordinate directions. The most common cause of such forces is gravity.

While inertia forces are rarely significant in problems of stress analysis in solids, they can only be neglected in fluids if the flows are sufficiently 'slow' to be dominated by pressure and viscous forces. In general, equations 2.7 to 2.9 would have to be modified to include fluid inertia forces. For example, equation 2.7 would become

$$\frac{\partial \sigma_{xx}}{\partial x} + \frac{\partial \sigma_{xy}}{\partial y} + \frac{\partial \sigma_{xz}}{\partial z} + \bar{X} = \rho\left(\frac{\partial u}{\partial t} + u\frac{\partial u}{\partial x} + v\frac{\partial u}{\partial y} + w\frac{\partial u}{\partial z}\right) \tag{2.10}$$

where t is time and ρ is the density of the fluid. This, together with the corresponding forms for the y- and z-directions, constitutes the Navier–Stokes equations.

2.1.3 Energy equation In the absence of significant thermal convection effects, the condition for conservation of energy within a solid or fluid medium may be expressed as

$$k\left(\frac{\partial^2 T}{\partial x^2} + \frac{\partial^2 T}{\partial y^2} + \frac{\partial^2 T}{\partial z^2}\right) + g = \rho C_p \frac{\partial T}{\partial t} \tag{2.11}$$

where T is temperature, g is the heat generated per unit volume, and k and C_p are the thermal conductivity and specific heat of the material. Heat generation may, for example, be due to the mechanical work performed on the material, in which case

$$g = \sigma_{xx}\,e_{xx} + \sigma_{yy}\,e_{yy} + \sigma_{zz}\,e_{zz} + \sigma_{xy}\,e_{xy} + \sigma_{yz}\,e_{yz} + \sigma_{zx}\,e_{zx} \qquad (2.12)$$

In the case of a fluid medium in which a significant amount of heat is transferred by convection, the following expression would have to be added to the right-hand side of equation 2.11

$$\rho C_p \left(u\,\frac{\partial T}{\partial x} + v\,\frac{\partial T}{\partial y} + w\,\frac{\partial T}{\partial z} \right) \qquad (2.13)$$

2.1.4 Compatibility equations   Strains or strain rates must be compatible with each other. The physical interpretation of compatibility is that no discontinuities such as holes or overlaps of material exist in the continuum: u, v and w are continuous and differentiable functions of position. Hence, from the six strain (rate) definitions given in equations 2.2 to 2.5, the following six relationships can be obtained by eliminating u, v and w by differentiation in various ways.

$$\frac{\partial^2 e_{xx}}{\partial y^2} + \frac{\partial^2 e_{yy}}{\partial x^2} = \frac{\partial^2 e_{xy}}{\partial x\,\partial y} \qquad (2.14)$$

$$\frac{\partial^2 e_{yy}}{\partial z^2} + \frac{\partial^2 e_{zz}}{\partial y^2} = \frac{\partial^2 e_{yz}}{\partial y\,\partial z} \qquad (2.15)$$

$$\frac{\partial^2 e_{zz}}{\partial x^2} + \frac{\partial^2 e_{xx}}{\partial z^2} = \frac{\partial^2 e_{zx}}{\partial z\,\partial x} \qquad (2.16)$$

$$2\,\frac{\partial^2 e_{xx}}{\partial y\,\partial z} = \frac{\partial}{\partial x}\left(-\frac{\partial e_{yz}}{\partial x} + \frac{\partial e_{zx}}{\partial y} + \frac{\partial e_{xy}}{\partial z} \right) \qquad (2.17)$$

$$2\,\frac{\partial^2 e_{yy}}{\partial z\,\partial x} = \frac{\partial}{\partial y}\left(\frac{\partial e_{yz}}{\partial x} - \frac{\partial e_{zx}}{\partial y} + \frac{\partial e_{xy}}{\partial z} \right) \qquad (2.18)$$

$$2\,\frac{\partial^2 e_{zz}}{\partial x\,\partial y} = \frac{\partial}{\partial z}\left(\frac{\partial e_{yz}}{\partial x} + \frac{\partial e_{zx}}{\partial y} - \frac{\partial e_{xy}}{\partial z} \right) \qquad (2.19)$$

These compatibility equations are rarely stated explicitly in fluid mechanics analyses, because such analyses are normally formulated with velocities as the unknowns. Compatibility of strain rates is thereby automatically satisfied. On the other hand, many problems of stress analysis in solids are formulated with stresses as the unknowns, when care must be taken to ensure that the strains obtained are compatible. These points are illustrated by the practical examples considered in section 2.2.

2.1.5 Continuity equation   The continuity equation for fluid flow expresses the condition for conservation of mass. For a fluid with constant density it reduces to the incompressibility condition

$$e_{xx} + e_{yy} + e_{zz} = 0 \qquad (2.20)$$

Volume changes in solids are accounted for by the constitutive relationships.

2.1.6 Constitutive equations   The relationships between stresses and strains or strain rates are expressed in terms of constitutive equations which introduce the relevant material properties. For present purposes the existence of viscoelastic materials is ignored, and solids and fluids are assumed to be purely elastic and purely viscous respectively.

For an elastic solid the strains defined in equations 2.2 to 2.5 may be produced both by the application of stress and by raising the temperature. Although thermally induced strains are small, so by assumption are the elastic strains. In general the properties of the material may vary. If they are independent of position within the body, direction at any particular point, and stress or strain applied, the material is said to be homogeneous, isotropic and linearly elastic, and the constitutive equations are

$$e_{xx} = \frac{1}{E} \left[\sigma_{xx} - \nu(\sigma_{yy} + \sigma_{zz}) \right] + \alpha \, \Delta T \qquad (2.21)$$

$$e_{yy} = \frac{1}{E} \left[\sigma_{yy} - \nu(\sigma_{zz} + \sigma_{xx}) \right] + \alpha \, \Delta T \qquad (2.22)$$

$$e_{zz} = \frac{1}{E} \left[\sigma_{zz} - \nu(\sigma_{xx} + \sigma_{yy}) \right] + \alpha \, \Delta T \qquad (2.23)$$

$$e_{xy} = \frac{\sigma_{xy}}{G} = \frac{2(1+\nu)}{E} \sigma_{xy} \qquad (2.24)$$

$$e_{yz} = \frac{\sigma_{yz}}{G} = \frac{2(1+\nu)}{E} \sigma_{yz} \qquad (2.25)$$

$$e_{zx} = \frac{\sigma_{zx}}{G} = \frac{2(1+\nu)}{E} \sigma_{zx} \qquad (2.26)$$

where E is Young's modulus, G is the shear modulus, ν is Poisson's ratio, α is the coefficient of thermal expansion and ΔT is the temperature rise. In the absence of thermal strains, the volumetric strain is

$$e_{xx} + e_{yy} + e_{zz} = \frac{(1 - 2\nu)}{E} (\sigma_{xx} + \sigma_{yy} + \sigma_{zz}) \qquad (2.27)$$

Clearly, if $\nu = \frac{1}{2}$ the material is incompressible.

For a newtonian fluid, which by definition is homogeneous, isotropic and linear, undergoing laminar flow the constitutive equations are

$$\sigma_{xx} = -p + 2\mu e_{xx}, \sigma_{yy} = -p + 2\mu e_{yy}, \sigma_{zz} = -p + 2\mu e_{zz} \qquad (2.28)$$

$$\sigma_{xy} = \mu e_{xy}, \sigma_{yz} = \mu e_{yz}, \sigma_{zx} = \mu e_{zx} \qquad (2.29)$$

where μ is the viscosity and p is the hydrostatic pressure.

It is worth noting that constitutive equations for solids usually express strains as functions of stresses, whereas for fluids stresses are expressed as functions of strain rates. These arrangements reflect the usual choices of unknowns discussed in section 2.1.4.

2.2 Some Physical Problems

The following examples serve to demonstrate the application of the fundamental equations of continuum mechanics to practical problems. The reader may not be familiar with the details of all of these examples, but should note the similarities between the resulting differential equations.

2.2.1 Downstream viscous flow in a uniform channel   Figure 2.2 shows the cross-section of a uniform channel, in this case rectangular in shape. The lower three sides of the channel are stationary while the top boundary moves with a velocity V_z in the z-direction normal to the cross-section. The velocity of flow, w, is also in this direction. Since the channel is uniform there are no variations with z, with the exception that pressure is a linear function of z. Substituting the expressions for stresses given in equations 2.28 and 2.29 into equilibrium equation 2.9 and neglecting body forces, the governing differential equation for w is obtained as

$$\frac{\partial}{\partial x}\left(\mu \frac{\partial w}{\partial x}\right) + \frac{\partial}{\partial y}\left(\mu \frac{\partial w}{\partial y}\right) = \frac{\partial p}{\partial z} \equiv P_z \qquad (2.30)$$

Since the viscosity is constant

$$\frac{\partial^2 w}{\partial x^2} + \frac{\partial^2 w}{\partial y^2} \equiv \nabla^2 w = \frac{P_z}{\mu} \qquad (2.31)$$

where ∇^2 is the harmonic operator. The fact that the pressure gradients in the x- and y-directions (in this case both are zero) do not vary in the z-direction means that the downstream pressure gradient P_z is independent of x and y, because

$$\frac{\partial P_z}{\partial x} = \frac{\partial}{\partial x}\left(\frac{\partial p}{\partial z}\right) = \frac{\partial}{\partial z}\left(\frac{\partial p}{\partial x}\right) = 0 \qquad (2.32)$$

$$\frac{\partial P_z}{\partial y} = \frac{\partial}{\partial y}\left(\frac{\partial p}{\partial z}\right) = \frac{\partial}{\partial z}\left(\frac{\partial p}{\partial y}\right) = 0 \qquad (2.33)$$

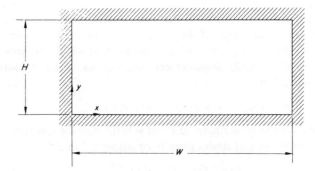

Figure 2.2 Rectangular channel geometry and co-ordinates

Assuming there is no slip between the fluid and the solid walls of the channel, the boundary conditions take the form of prescribed velocities

$$w = 0 \text{ on } x = 0, x = W \text{ and } y = 0; \quad w = V_z \text{ on } y = H \tag{2.34}$$

The volumetric downstream flow rate can be found by integration of the velocity profile over the cross-section as

$$Q = \int_0^H \int_0^W w \, dx \, dy \tag{2.35}$$

2.2.2 Torsion of a prismatic bar Figure 2.3 shows the cross-section of a prismatic bar, in this case elliptical in shape. The bar is twisted about the z-axis normal to the cross-section. According to St Venant's theory of torsion (see, for example, Ford, 1963) the deformation of the bar is composed of a rotation of the cross-section and a warping in the z-direction, although there is no warping

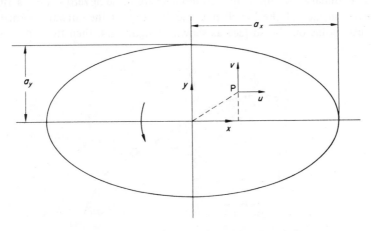

Figure 2.3 Elliptical cross-section of a prismatic bar

in the special case of a cylindrical bar. The anticlockwise angle of rotation is taken to be θ per unit length of the bar and the cross-section shown is taken to be a distance z from the position of zero rotation, at which the axes are fixed. Consequently, the (small) displacement components of the typical point P shown in figure 2.3 are

$$u = -\theta z y, \qquad v = \theta z x, \qquad w = w(x,y) \qquad (2.36)$$

where x and y are the co-ordinates of P and w is the warping function.

From the definitions of strains given in equations 2.2 to 2.5

$$e_{xx} = e_{yy} = e_{zz} = e_{xy} = 0$$

$$e_{yz} = \theta x + \frac{\partial w}{\partial y}, \qquad e_{zx} = \frac{\partial w}{\partial x} - \theta y \qquad (2.37)$$

The compatibility equations 2.14 to 2.19 are automatically satisfied because the problem is formulated with displacements as the unknowns. From the constitutive equations 2.21 to 2.26, at constant temperature the only nonzero stress components are $\sigma_{yz} = G e_{yz}$ and $\sigma_{zx} = G e_{zx}$. Of the equilibrium conditions, only equation 2.9 is relevant, and in the absence of body forces becomes

$$\frac{\partial \sigma_{zx}}{\partial x} + \frac{\partial \sigma_{zy}}{\partial y} = 0 \qquad (2.38)$$

Hence

$$\frac{\partial}{\partial x}\left[G\left(\frac{\partial w}{\partial x} - \theta y\right) \right] + \frac{\partial}{\partial y}\left[G\left(\theta x + \frac{\partial w}{\partial y}\right) \right] = 0$$

$$\frac{\partial^2 w}{\partial x^2} + \frac{\partial^2 w}{\partial y^2} \equiv \nabla^2 w = 0 \qquad (2.39)$$

The boundary conditions for this problem are those of zero stresses acting on the outer surface of the bar. If n is the direction of the outward normal at a particular point on the surface as shown in figure 2.4, then the surface stress

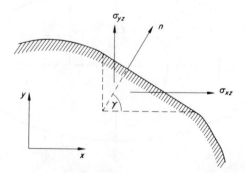

Figure 2.4   Part of the boundary of a cross-section of a prismatic bar

component σ_{zn} is zero, as is the complementary stress σ_{nz}. Hence

$$\sigma_{nz} = \sigma_{xz} \cos \gamma + \sigma_{yz} \sin \gamma = 0 \qquad (2.40)$$

$$\left(\frac{\partial w}{\partial x} - \theta y\right) \cos \gamma + \left(\frac{\partial w}{\partial y} + \theta x\right) \sin \gamma = 0 \qquad (2.41)$$

where γ is the angle between the normal and the x-axis. Boundary conditions for w of this type are difficult to apply, particularly in the case of asymmetrical bars where the position of the axis of rotation is not immediately apparent.

This example provides a good illustration of the disadvantages of using displacements as the unknowns in a problem whose boundary conditions involve stresses. Because of the difficulty of applying these conditions, solid body problems are often formulated in terms of stress functions. In the present problem a stress function χ can be defined which automatically satisfies the equilibrium equation 2.38

$$\sigma_{zx} = \frac{\partial \chi}{\partial y}, \quad \sigma_{yz} = -\frac{\partial \chi}{\partial x} \qquad (2.42)$$

The strains are therefore given by

$$e_{zx} = \frac{1}{G} \frac{\partial \chi}{\partial y}, \quad e_{yz} = -\frac{1}{G} \frac{\partial \chi}{\partial x} \qquad (2.43)$$

and these must satisfy compatibility equations 2.17 and 2.18 which require that

$$\frac{\partial e_{zx}}{\partial y} - \frac{\partial e_{yz}}{\partial x} = \text{constant} \qquad (2.44)$$

the value of the constant being obtained from equations 2.37 as -2θ. Hence, substituting the strains defined in equations 2.43 into equation 2.44, the governing differential equation for χ is obtained as

$$\frac{\partial^2 \chi}{\partial x^2} + \frac{\partial^2 \chi}{\partial y^2} = \nabla^2 \chi = -2G\theta \qquad (2.45)$$

whose mathematical form is very similar to that of equation 2.39. The boundary condition $\sigma_{nz} = 0$ is obtained when the gradient of χ along the boundary is zero, that is, when the value of χ is constant along the boundary. Since it is not the absolute value of χ but only its derivatives which determine the stress distribution, $\chi = 0$ is a suitable boundary condition, and one that is much easier to apply than equation 2.41 for w.

Having obtained the stress distributions, the magnitude of the couple required to twist the bar may be found by integration over the cross-section

$$C = \int \int (x\sigma_{yz} - y\sigma_{xz}) \, dx \, dy \qquad (2.46)$$

Introducing the stress function

$$C = - \int \int \left(x \frac{\partial \chi}{\partial x} + y \frac{\partial \chi}{\partial y} \right) dx\, dy$$

$$= - \int \int \left[\frac{\partial}{\partial x} (x\chi) + \frac{\partial}{\partial y} (y\chi) \right] dx\, dy + 2 \int \int \chi\, dx\, dy$$

Applying Green's theorem to the first of these integrals

$$C = - \oint (x\chi\, dy - y\chi\, dx) + 2 \int \int \chi\, dx\, dy$$

where the line integration is performed around the boundary of the bar. Since $\chi = 0$ on this boundary

$$C = 2 \int \int \chi\, dx\, dy \qquad (2.47)$$

2.2.3 Ideal fluid flow   The 'ideal' fluid model is often used to describe the motion of real fluids in regions remote from wakes or boundary layers formed near solid boundaries. In these latter regions viscous effects are significant, whereas an ideal fluid, in addition to being homogeneous, isotropic and incompressible, is also inviscid. The flow of such a fluid is governed by pressure and inertia forces. For two-dimensional flow in the $x-y$ plane (that is, with $w = 0$ everywhere) the equilibrium conditions in the form of equation 2.10 for steady flow in which body forces are negligible reduce to

$$-\frac{\partial p}{\partial x} = \rho \left(u \frac{\partial u}{\partial x} + v \frac{\partial u}{\partial y} \right) \qquad (2.48)$$

$$-\frac{\partial p}{\partial y} = \rho \left(u \frac{\partial v}{\partial x} + v \frac{\partial v}{\partial y} \right) \qquad (2.49)$$

The two-dimensional form of the continuity equation 2.20 is

$$\frac{\partial u}{\partial x} + \frac{\partial v}{\partial y} = 0 \qquad (2.50)$$

It is convenient to define a stream function ψ which automatically satisfies continuity

$$u = \frac{\partial \psi}{\partial y}, \qquad v = -\frac{\partial \psi}{\partial x} \qquad (2.51)$$

Rearranging equations 2.48 and 2.49

$$-\frac{\partial}{\partial x} \left(\frac{p}{\rho} + \frac{u^2 + v^2}{2} \right) = v \left(\frac{\partial u}{\partial y} - \frac{\partial v}{\partial x} \right) \qquad (2.52)$$

$$-\frac{\partial}{\partial y}\left(\frac{p}{\rho}+\frac{u^2+v^2}{2}\right)=u\left(\frac{\partial v}{\partial x}-\frac{\partial u}{\partial y}\right) \tag{2.53}$$

which in general will only both be satisfied if

$$\frac{\partial u}{\partial y}-\frac{\partial v}{\partial x}=0 \tag{2.54}$$

This condition implies that the flow is 'irrotational' and is a consequence of neglecting viscous effects. Using the definitions for velocities in terms of stream function given in equations 2.51, equation 2.54 becomes

$$\nabla^2\psi=0 \tag{2.55}$$

Once the stream function distribution has been found, the pressure distribution can be determined from

$$\frac{p}{\rho}+\frac{u^2+v^2}{2}=\text{constant} \tag{2.56}$$

2.2.4 Diffusion problems   There are many physical processes which involve some form of diffusion. An example is the conduction of heat in solid or fluid media, which is governed by the energy equation 2.11. For steady conduction in the $x-y$ plane

$$\nabla^2 T=-g/k \tag{2.57}$$

Various types of boundary conditions may be appropriate. The simplest is where the temperature at the boundary or part of the boundary is prescribed. A thermally insulated boundary gives rise to the derivative condition

$$\frac{\partial T}{\partial n}=0 \tag{2.58}$$

where n is the direction of the outward normal to the boundary. A more general type of condition is obtained when the heat conducted to the boundary is convected away in a surrounding fluid medium

$$-k\frac{\partial T}{\partial n}=h(T-T_\infty) \tag{2.59}$$

where T_∞ is the bulk temperature of the medium and h is the heat transfer coefficient.

Further examples of diffusion processes are provided by electrical conduction, electrostatic-potential distribution in insulating media, fluid flow in porous media, and neutron diffusion in a nuclear reactor. Consider, for example, the flow of fluid in a porous medium, such as the seepage flow of water in soil. For a newtonian flow, the volumetric flow rate along an individual passage in the porous medium is proportional to the pressure gradient there. Taking a macroscopic view of the flow, the mean velocity in any direction (effectively the sum of the flow rates along individual passages) is also proportional to the pressure gradient in that direction. Hence, for two-dimensional flow in the $x-y$ plane

$$\bar{u} = -\kappa \frac{\partial p}{\partial x}, \quad \bar{v} = -\kappa \frac{\partial p}{\partial y} \tag{2.60}$$

where κ is the permeability of the porous medium. As with ideal fluid flow discussed in the last subsection, a stream function ψ may be defined which automatically satisfies continuity

$$\bar{u} = \frac{\partial \psi}{\partial y}, \quad \bar{v} = -\frac{\partial \psi}{\partial x} \tag{2.61}$$

Eliminating pressure from equations 2.60 by differentiation, the governing equation becomes

$$\frac{\partial \bar{u}}{\partial y} - \frac{\partial \bar{v}}{\partial x} = \nabla^2 \psi = 0 \tag{2.62}$$

Boundary conditions are generally of the form of either prescribed constant values of ψ for boundaries impervious to flow, or zero derivatives normal to boundaries where the pressure is constant.

2.2.5 Plane strain   Both plane strain and plane stress, which is considered in the next subsection, are important modes of two-dimensional solid deformation, to which a considerable amount of attention is devoted in later chapters. Figure 2.5 shows a solid body whose cross-section is uniform in the z-direction. Provided its length in this direction is large, the typical section OABC can be regarded as being remote from the ends. Assuming the surface tractions applied to the body are in the $x-y$ plane, the resulting state of strain at such a section is two-dimensional, being independent of z and with $w = 0$.

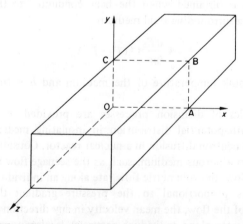

Figure 2.5   The plane strain approximation for a prismatic solid body

In the absence of body forces, the equilibrium equations 2.7 to 2.9 reduce to

$$\frac{\partial \sigma_{xx}}{\partial x} + \frac{\partial \sigma_{xy}}{\partial y} = 0 \tag{2.63}$$

$$\frac{\partial \sigma_{yx}}{\partial x} + \frac{\partial \sigma_{yy}}{\partial y} = 0 \tag{2.64}$$

$$\frac{\partial \sigma_{zz}}{\partial z} = 0 \tag{2.65}$$

From the strain definitions, equations 2.2 to 2.5, $e_{zz} = e_{yz} = e_{zx} = 0$ and from equation 2.23

$$\sigma_{zz} = \nu(\sigma_{xx} + \sigma_{yy}) \tag{2.66}$$

provided there are no temperature changes.

The usual method of proceeding with this analysis is to introduce a stress function χ, in this case one often known as Airy's stress function, which automatically satisfies equilibrium equations 2.63 and 2.64

$$\sigma_{xx} = \frac{\partial^2 \chi}{\partial y^2}, \quad \sigma_{yy} = \frac{\partial^2 \chi}{\partial x^2}, \quad \sigma_{xy} = -\frac{\partial^2 \chi}{\partial x \partial y} \tag{2.67}$$

Hence, from equation 2.66

$$\sigma_{zz} = \nu \nabla^2 \chi \tag{2.68}$$

and the nonzero strain components may be obtained from equations 2.21, 2.22 and 2.24 as

$$e_{xx} = \frac{1}{E}\left[\frac{\partial^2 \chi}{\partial y^2} - \nu\left(\frac{\partial^2 \chi}{\partial x^2} + \nu \nabla^2 \chi\right)\right] \tag{2.69}$$

$$e_{yy} = \frac{1}{E}\left[\frac{\partial^2 \chi}{\partial x^2} - \nu\left(\frac{\partial^2 \chi}{\partial y^2} + \nu \nabla^2 \chi\right)\right] \tag{2.70}$$

$$e_{xy} = -\frac{2(1 + \nu)}{E} \frac{\partial^2 \chi}{\partial x \partial y} \tag{2.71}$$

Now it remains to ensure that these strains are compatible. Equations 2.15 to 2.19 are automatically satisfied by the plane strain assumptions and substitution into equation 2.14 yields

$$\frac{\partial^4 \chi}{\partial x^4} + \frac{\partial^4 \chi}{\partial y^4} - \nu\left[2\frac{\partial^4 \chi}{\partial x^2 \partial y^2} + \nu \nabla^2 (\nabla^2 \chi)\right] = -2(1 + \nu)\frac{\partial^4 \chi}{\partial x^2 \partial y^2}$$

which reduces to

$$\nabla^4 \chi = 0 \tag{2.72}$$

for all values of ν, the biharmonic operator ∇^4 being such that

$$\nabla^4 \chi \equiv \nabla^2(\nabla^2 \chi) \equiv \frac{\partial^4 \chi}{\partial x^4} + \frac{\partial^4 \chi}{\partial y^4} + 2\frac{\partial^4 \chi}{\partial x^2 \partial y^2} \tag{2.73}$$

Boundary conditions for governing equation 2.72 may involve either prescribed stresses or displacements. Alternatively, a boundary might be constrained to move freely in a prescribed direction. Practical examples of such conditions are given in later chapters.

2.2.6 Plane stress   Figure 2.6 shows a solid plate lying in the x–y plane. If the applied tractions are in the same plane the stresses on the faces of the plate are zero. Provided the plate is sufficiently thin the plane stress approximation $\sigma_{zz} = \sigma_{yz} = \sigma_{xz} = 0$ can be applied throughout the material. In the absence of body forces, the equilibrium conditions again reduce to equations 2.63 and 2.64, permitting the use of Airy's stress function defined by equation 2.67.

From the constitutive equations 2.21 to 2.26, $e_{yz} = e_{zx} = 0$ and in the absence of temperature changes

$$e_{xx} = \frac{1}{E}(\sigma_{xx} - \nu\sigma_{yy}) = \frac{1}{E}\left(\frac{\partial^2 \chi}{\partial y^2} - \nu\frac{\partial^2 \chi}{\partial x^2}\right) \tag{2.74}$$

$$e_{yy} = \frac{1}{E}(\sigma_{yy} - \nu\sigma_{xx}) = \frac{1}{E}\left(\frac{\partial^2 \chi}{\partial x^2} - \nu\frac{\partial^2 \chi}{\partial y^2}\right) \tag{2.75}$$

$$e_{zz} = -\frac{\nu}{E}(\sigma_{xx} + \sigma_{yy}) = -\frac{\nu}{E}\nabla^2 \chi \tag{2.76}$$

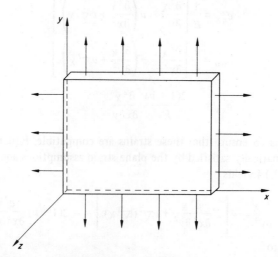

Figure 2.6   The plane stress approximation for a thin solid body

$$e_{xy} = \frac{2(1 + \nu)}{E} \sigma_{xy} = - \frac{2(1 + \nu)}{E} \frac{\partial^2 \chi}{\partial x \partial y} \tag{2.77}$$

Substituting these expressions for strains into compatibility equation 2.14

$$\frac{\partial^4 \chi}{\partial x^4} + \frac{\partial^4 \chi}{\partial y^4} - 2\nu \frac{\partial^4 \chi}{\partial x^2 \partial y^2} = - 2(1 + \nu) \frac{\partial^4 \chi}{\partial x^2 \partial y^2}$$

$$\nabla^4 \chi = 0 \tag{2.78}$$

for all values of ν. Although the remaining compatibility equations 2.15 to 2.19 are not automatically satisfied, it can be shown that they are satisfied approximately if the plate is thin enough for variations of stresses and strains through the thickness to be neglected. Boundary conditions for plane stress problems may be of the stress or displacement types described for plane strain.

2.2.7 Recirculating viscous flow The equation governing downstream viscous flow in a uniform channel such as that shown in figure 2.2 was derived in section 2.2.1. If the upper boundary of this channel moves with a velocity component V_x in the x-direction relative to the remaining boundaries which are stationary, a recirculating viscous flow in the plane of the cross-section is created. It is convenient to define a stream function as in equations 2.51 which automatically satisfies continuity for this two-dimensional flow. In the absence of body forces the equilibrium conditions are given by equations 2.63 and 2.64. Using constitutive equations 2.28 and 2.29, these may be expressed in terms of pressure and velocity gradients as

$$P_x \equiv \frac{\partial p}{\partial x} = \mu \left[2 \frac{\partial^2 u}{\partial x^2} + \frac{\partial}{\partial y} \left(\frac{\partial u}{\partial y} + \frac{\partial v}{\partial x} \right) \right] \tag{2.79}$$

$$P_y \equiv \frac{\partial p}{\partial y} = \mu \left[2 \frac{\partial^2 v}{\partial y^2} + \frac{\partial}{\partial x} \left(\frac{\partial u}{\partial y} + \frac{\partial v}{\partial x} \right) \right] \tag{2.80}$$

Eliminating pressure from these equations by differentiation, and introducing the stream function, the following governing equation is obtained

$$\nabla^4 \psi = 0 \tag{2.81}$$

Assuming there is no slip between the fluid and the solid walls of the channel, the boundary conditions for the problem illustrated in figure 2.2 are

$$\psi = 0, \quad v = -\frac{\partial \psi}{\partial x} = 0 \quad \text{on } x = 0, x = W$$

$$\psi = 0, \quad u = \frac{\partial \psi}{\partial y} = 0 \quad \text{on } y = 0$$

$$\psi = Q_L, \quad u = \frac{\partial \psi}{\partial y} = V_x \quad \text{on } y = H \tag{2.82}$$

where Q_L is the volumetric flow rate leaked into and out of the channel between the moving and fixed boundaries. The absolute values of the stream function prescribed on these boundaries are unimportant, provided a difference of Q_L is maintained between them: the velocity profiles depend only on the derivatives of ψ.

In the present problem of recirculating viscous flow, the direct strain rate in the z-direction is zero. Consequently there is a close analogy with the plane strain deformation of solid bodies outlined in section 2.2.5.

2.2.8 Laterally loaded flat plate   The final example concerns a flat plate, thickness h, lying in the $x-y$ plane and subjected to a lateral pressure $p(x, y)$. The equation governing the deflection w may be stated without proof as

$$\nabla^4 w = p/D \tag{2.83}$$

where $D = Eh^3/12(1 - \nu^2)$ is the flexural rigidity. Possible boundary conditions include $w = 0$ for a supported edge, and $\partial w/\partial n = 0$ (n being the direction of the outward normal in the $x-y$ plane at the boundary) for an edge of the plate which is clamped. While such plate bending problems are outside the scope of this book, it is interesting to note the familiar form of the governing equation.

2.2.9 General comments   Apart from the similarities between the governing partial differential equations, which are considered in the next section, there are a number of general comments to be made about the foregoing examples. Solid body problems are often formulated in terms of stress functions which automatically satisfy equilibrium, and solutions are obtained which satisfy compatibility of strains. On the other hand, fluid flow problems are usually formulated in terms of stream functions which automatically satisfy continuity (incompressibility), and solutions are obtained which satisfy equilibrium. The main reason for these strategies is one of convenience in view of the nature of the boundary conditions: stresses and velocities are generally prescribed for solids and fluids respectively. Such formulations are not necessarily appropriate when using finite element methods, particularly when more general boundary conditions of mixed types are to be applied. It should be noted that both stress function and stream function formulations cannot be readily applied to three-dimensional problems, although these are outside the scope of this book.

Boundary conditions often prescribe either the value of the dependent variable on a boundary, or the value of its first derivative normal to the boundary. A more general condition, of which the foregoing are special cases, is provided by equation 2.59 which is of the form

$$a_1 \frac{\partial \psi}{\partial n} + a_2 \psi + a_3 = 0 \tag{2.84}$$

where ψ is the dependent variable and a_1, a_2 and a_3 are prescribed constants. The number of boundary conditions required is determined by the order of the governing differential equation. The second-order equations arising in sections 2.2.1 to 2.2.4 require two conditions per co-ordinate direction, one on each boundary. The fourth-order equations arising in sections 2.2.5 to 2.2.8 require four conditions, two on each boundary.

The principal result of solving the governing equations with the relevant boundary conditions for a particular problem is to obtain the dependent variable as a function of position within the solution domain. Some problems call for further analysis. For example, flow rate in section 2.2.1 and torsional couple in section 2.2.2 may be obtained by integration of the dependent variables over the solution domains. While in many of the problems described, body forces and thermal strains are assumed to be negligible, their inclusion would have no effect on the fundamental types of resulting differential equations. It is these types which are important in determining the methods of solution to be employed.

2.3   Classification of Partial Differential Equations

The general form of a second-order partial differential equation involving two independent variables may be stated as

$$A \frac{\partial^2 \psi}{\partial x^2} + B \frac{\partial^2 \psi}{\partial x \partial y} + C \frac{\partial^2 \psi}{\partial y^2} + D \frac{\partial \psi}{\partial x} + E \frac{\partial \psi}{\partial y} + F\psi + G = 0 \qquad (2.85)$$

where ψ is the dependent variable and x and y are the independent variables. If the coefficients A to G are functions of x and y only, the equation is said to be linear. In nonlinear equations they also depend on ψ or its derivatives. As shown by, for example, Crandall (1956), the values of these coefficients determine the type of equation, and hence the method of solution. The important parameter is

$$\lambda = B^2 - 4AC \qquad (2.86)$$

and equation 2.85 is said to be elliptic, parabolic or hyperbolic according to whether λ is negative, zero or positive. While it is possible for the type of an equation to change within the solution domain if A, B or C vary, in the majority of practical problems this does not happen. Similar classifications can be applied to higher-order equations and to those involving more than two independent variables.

Elliptic equations normally occur in equilibrium problems, whereas the parabolic and hyperbolic types occur in propagation problems. A distinction between equilibrium and propagation problems can be made in terms of the type of conditions applied at the boundaries of the solution domain (see, for example, Crandall (1956) or Fenner (1974)). The domain for an equilibrium problem is closed and boundary conditions are prescribed around the entire boundary: such a problem is often said to be of the boundary-value type. In this

book, attention is concentrated on finite element methods for solving
boundary-value problems, although such methods can also be applied to
propagation problems.

2.3.1 Harmonic and biharmonic equations   The governing equations for the
problems outlined in sections 2.2 are essentially of two types.

$$\frac{\partial^2 \psi}{\partial x^2} + \frac{\partial^2 \psi}{\partial y^2} = \nabla^2 \psi = \phi_1(x, y) \tag{2.87}$$

$$\frac{\partial^4 \psi}{\partial x^4} + \frac{\partial^4 \psi}{\partial y^4} + 2\frac{\partial^4 \psi}{\partial x^2 \partial y^2} = \nabla^2(\nabla^2 \psi) = \nabla^4 \psi = \phi_2(x, y) \tag{2.88}$$

Equation 2.87 is elliptic according to the above definition (since $\lambda = -4$), and is
often referred to as Poisson's equation. Laplace's equation is obtained as a
special case when $\phi_1 = 0$. The fourth-order equation 2.88 is also elliptic. It is
convenient to distinguish between the two types by referring to equation 2.87 as
being harmonic, and to equation 2.88 as being biharmonic.

Of the problems considered in section 2.2, the following are governed by
harmonic equations: downstream viscous flow, torsion, ideal fluid flow and
diffusion processes. The remainder, namely plane strain, plane stress, recirculating viscous flow and the bending of flat plates, are governed by biharmonic
equations. That all of these problems are of the boundary-value (elliptic) type is
due to the fact that they involve either static stresses in solids or steady fluid
flows. If, for example, the time-dependent term in equation 2.11 is retained in
equation 2.57 the resulting unsteady conduction problem is of the propagation
type.

2.3.2 Quasi-harmonic equations   In addition to problems governed by the
harmonic or biharmonic equations 2.87 or 2.88, there are many others which
give rise to equations which are of essentially the same types, and which can be
solved by very similar methods. A good example is provided by hydrodynamic
lubrication in a thin film between moving bearing surfaces. If V_x is the relative
velocity, in the x-direction, between surfaces which are parallel to the $x-y$
plane, the pressure in the lubricant is governed by

$$\frac{\partial}{\partial x}\left(H^3 \frac{\partial p}{\partial x}\right) + \frac{\partial}{\partial y}\left(H^3 \frac{\partial p}{\partial y}\right) = 6\mu V_x \frac{\partial H}{\partial x} \tag{2.89}$$

where μ is the viscosity and $H(x, y)$ is the thickness of the film. This equation is
linear in the sense defined in connection with equation 2.85, and is
quasi-harmonic in that but for the variation of H it would conform to equation
2.87. Another (nonlinear) example is provided by equation 2.30 if the viscosity
is non-newtonian and depends on the local velocity gradients.

2.4   Methods for Solving Harmonic and Biharmonic Equations

With the exception of a few simple cases, harmonic and biharmonic partial differential equations cannot be solved analytically. While there are many mathematical functions that satisfy the differential equations, they rarely satisfy the required boundary conditions: in boundary-value problems it is the boundary conditions as much as the differential equations that determine the solution. One method of solution, however, is to employ an infinite series of such functions to satisfy the boundary conditions, although such an approach may involve as much computation as purely numerical methods.

The principal numerical methods of solution involve discretisation: the continuous functions (such as stresses, displacements or velocities) are represented approximately by values at a finite number of points within the solution domain. These values are obtained from sets of simultaneous (and usually linear) algebraic equations. The accuracy of the representations increases with the number of points used, particularly if they are concentrated in regions where the functions vary most rapidly. Various methods of analysis are used to assemble the relevant sets of algebraic equations, the main ones being finite element and finite difference methods. While the former are the subject of this book, it is worth briefly reviewing finite difference methods in order to be able to compare the two approaches.

2.4.1 The finite difference approach   Finite difference methods involve replacing derivatives by difference approximations. Figure 2.7 shows part of a grid in a two-dimensional solution domain in the $x-y$ plane. The lines of the grid are uniformly spaced in the x- and y-directions, the distances between them being h_x and h_y respectively. The points to be used in the finite difference analysis are located at the intersections of the grid lines. The point labelled O may be

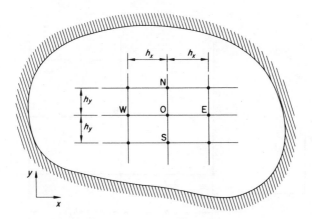

Figure 2.7   Rectangular grid in a two-dimensional solution domain

regarded as a typical grid point within the solution domain, and the compass-point labels N, S, E and W are used for the four adjacent grid points. The derivative with respect to x in equation 2.87 at the point O may be expressed in terms of the values of ψ at the grid points by means of the central difference formula

$$\left(\frac{\partial^2 \psi}{\partial x^2}\right)_O = \frac{\psi_W - 2\psi_O + \psi_E}{h_x^2} + e_T \qquad (2.90)$$

This result may be obtained from Taylor's series, and the truncation error is of the order of

$$e_T \approx -\frac{h_x^2}{12}\left(\frac{\partial^4 \psi}{\partial x^4}\right)_O \qquad (2.91)$$

Using a similar expression for the derivative with respect to y, the following finite difference approximation for equation 2.87 is obtained

$$\frac{\psi_W - 2\psi_O + \psi_E}{h_x^2} + \frac{\psi_S - 2\psi_O + \psi_N}{h_y^2} = (\phi_1)_O \qquad (2.92)$$

which is applicable to all internal points.

Equations for the boundary points may be obtained from the relevant boundary conditions. For example, if the value of ψ at the point A shown in figure 2.8 is prescribed as α, then $\psi_A = \alpha$. Derivative boundary conditions can be applied with the aid of the relevant finite difference formulae. For example

$$\left(\frac{\partial \psi}{\partial x}\right)_A = \frac{\psi_B - \psi_A}{h_x} + O(h_x) \qquad (2.93)$$

and if this derivative is required to take the value β, then ψ_A is given by the

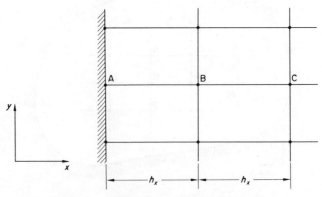

Figure 2.8   Grid points near a domain boundary

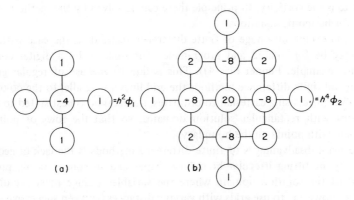

Figure 2.9 Computing molecules: (a) harmonic; (b) biharmonic

approximation

$$\psi_A = \psi_B - \beta h_x \qquad (2.94)$$

It should be noted, however, that the truncation error in equation 2.93 is of a different order from that in equation 2.90. A better approximation to the derivative at the boundary is provided by

$$\left(\frac{\partial \psi}{\partial x}\right)_A = \frac{1}{2h_x}(-3\psi_A + 4\psi_B - \psi_C) + O(h_x^2) \qquad (2.95)$$

hence

$$\psi_A = \frac{1}{3}(4\psi_B - \psi_C - 2\beta h_x) \qquad (2.96)$$

If there are n points in the solution domain then there are n linear equations for the values of ψ either of the form of equation 2.92, or expressing the boundary conditions.

For the special case $h_x = h_y = h$, equation 2.92 becomes

$$\psi_W + \psi_E + \psi_S + \psi_N - 4\psi_O = h^2(\phi_1)_O \qquad (2.97)$$

and is sometimes expressed in a more pictorial computing 'molecule' form as shown in figure 2.9a. The equivalent molecule for the biharmonic equation 2.88 is shown in figure 2.9b. In this case special treatment is required for both the points actually on the boundary and those immediately adjacent to it: for example, points A and B in figure 2.8. In principle the two conditions prescribed for each boundary can be used for this purpose.

Another method of solving a biharmonic equation is to reduce it to a pair of simultaneous harmonic equations. For example, in the case of recirculating viscous flow, equation 2.81 is often replaced by

$$\nabla^2 \psi = \omega, \qquad \nabla^2 \omega = 0 \qquad (2.98)$$

where ω is the vorticity. In principle these can be solved by the method outlined above for harmonic equations.

An important advantage of finite difference methods is the ease with which they may be formulated and programmed for solution by a digital computer (see, for example, Fenner (1974)). This is due to the use of regular grids. In virtually all finite difference methods the grid lines are parallel to the co-ordinate axes. Therefore, using cartesian co-ordinates, the method is best suited to solving problems with rectangular solution domains, so that the rows of points can terminate with points on the boundary.

The main disadvantages of finite difference methods is the lack of geometric flexibility in fitting irregular boundary shapes and in concentrating points in regions of the solution domain where the variables change most rapidly. It is possible, however, to use grids with varying distances between successive rows of points, and to introduce more sophisticated co-ordinate systems to suit the shape of the solution domain (see, for example, Gosman *et al.* (1969)). If the domain does not conform to one of the major co-ordinate systems, however, the problem is often more readily solved by a finite element method.

2.4.2 The finite element approach   Whereas finite difference methods are applied to the governing partial differential equations for the problems concerned, finite element methods are more often formulated directly from the physical arguments used to derive such equations. Nevertheless, it is important to identify the mathematical type of problem, such as harmonic or biharmonic, in order to guide the choice of a finite element formulation. The formulation and computer program developed for one problem are normally directly applicable to another problem of the same mathematical type. It is for this reason that the terms harmonic and biharmonic are used in later chapters to distinguish types of problems, although the finite element methods employed do not necessarily use the differential equations explicitly.

The finite element approach to solving continuum mechanics problems involves first dividing the solution domain into small subregions or finite elements. Such elements are either material or spatial subregions according to whether the continuum is solid or fluid. Associated with each element are several nodal points at which adjacent elements are effectively linked together, and at which the values of the relevant variables are to be determined. An analysis based on physical arguments such as equilibrium and compatibility is used to derive a set of simultaneous algebraic equations for these nodal point values. As in the analytical treatments outlined in section 2.2, there is usually a choice of variables between those of the displacement (velocity) and stress types. Displacement variables are used almost exclusively in this book: the resulting finite element formulations can readily accommodate both displacement and stress boundary conditions.

3 Finite Element Analysis of Harmonic Problems

In this chapter the formulation of a finite element analysis for two-dimensional problems of the harmonic type outlined in chapter 2 is described. One-dimensional problems of this type are considered by Fenner (1974), and three-dimensional problems are discussed in section 8.3. Attention is concentrated on the simplest types of elements, namely ones triangular in shape over which the strains or strain rates are assumed to be constant. Such elements are often referred to as constant strain triangles (CST). There is a close similarity of approach between the present formulation and the one described in chapter 1 for the analysis of structures.

3.1 Derivation of the Element Stiffness Matrix

Figure 3.1 shows a two-dimensional solution domain divided into a mesh of small triangular subregions or finite elements. Such elements are either material or spatial subregions according to whether the continuum is solid or fluid. Associated with each element are three nodal points, at the corners of the triangle. An advantage of using triangular elements is that they can be made to fit any shape of domain boundary, provided that the boundary can be represented with sufficient accuracy by a series of short straight lines.

The method of analysis described in this section and section 3.2 involves the direct application of conditions for equilibrium, and is often referred to as a direct equilibrium formulation. Because the formulation is so direct it is convenient to describe it in terms of a particular physical problem of the harmonic type. The problem of downstream viscous flow described in section 2.2.1 provides a suitable choice and serves to emphasise the applicability of finite element methods to fluid mechanics problems. In section 3.4 the alternative variational formulation is described which uses the governing differential equation 2.31 explicitly, and therefore shows how the present analysis can be generalised to solve any problem of the harmonic type.

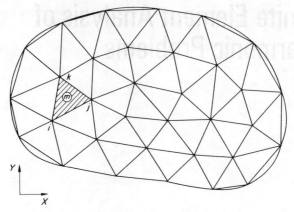

Figure 3.1   A two-dimensional solution domain divided into triangular finite elements

3.1.1 Element geometry and choice of shape function

All the elements and nodal points shown in figure 3.1 are numbered. The typical (shaded) element is numbered m and its nodes are numbered i, j and k. Figure 3.2 shows this element drawn to a larger scale. The cartesian co-ordinates X and Y shown in figure 3.1 are global co-ordinates for the overall solution domain, whereas x and y drawn in figure 3.2 are local to the particular element, with the origin at node i. The solution domain may be regarded as being of unit thickness in the z-direction, and the dimensions of the typical element in the plane of the

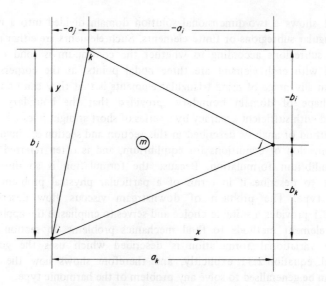

Figure 3.2   A typical triangular element

domain can be expressed in terms of the lengths a_i, a_j, a_k, b_i, b_j and b_k as shown. These may be obtained from the global co-ordinates of the nodal points as

$$
\begin{aligned}
a_i &= -X_j + X_k, & b_i &= Y_j - Y_k \\
a_j &= -X_k + X_i, & b_j &= Y_k - Y_i \\
a_k &= -X_i + X_j, & b_k &= Y_i - Y_j
\end{aligned}
\tag{3.1}
$$

Note that the subscripts in these definitions follow regular cyclic sequences. This set of dimensions is chosen for convenience in the subsequent analysis, in particular to give a regular form of coefficients in the dimension matrix defined in equation 3.12.

The area of the element can be calculated as the area of the enclosing rectangle, minus the areas of the three triangles outside the element. That is

$$
\Delta_m = a_k b_j - \tfrac{1}{2} a_i b_i + \tfrac{1}{2} a_j b_j + \tfrac{1}{2} a_k b_k
\tag{3.2}
$$

Since from figure 3.2

$$
\begin{aligned}
a_i + a_j + a_k &= 0 \\
b_i + b_j + b_k &= 0
\end{aligned}
\tag{3.3}
$$

equation 3.2 can be simplified by eliminating, say, a_i and b_i to give

$$
\Delta_m = \tfrac{1}{2}(a_k b_j - a_j b_k)
\tag{3.4}
$$

This expression is positive provided i, j and k are the numbers of the nodes of the element taken in anticlockwise order.

In the downstream viscous flow problem, the velocity in the z-direction, w, is the variable whose distribution over the solution domain is required. This velocity is assumed to vary linearly over each element

$$
w(x, y) = C_1 + C_2 x + C_3 y
\tag{3.5}
$$

where C_1, C_2 and C_3 are constant for the particular element. The expression defining the form of variation of the dependent variable over each element is often referred to as the shape function. Many forms of shape function are possible and some general types are discussed in section 8.2. Meanwhile equation 3.5 provides the simplest shape function suitable for the present type of problem and gives rise to constant rates of strain over each element. The use of a shape function involving three parameters C_1, C_2 and C_3 allows their values to be obtained in terms of the three nodal point velocities w_i, w_j and w_k.

$$
w_i = w(0, 0) = C_1
\tag{3.6}
$$

$$
w_j = w(a_k, -b_k) = C_1 + C_2 a_k - C_3 b_k
\tag{3.7}
$$

$$
w_k = w(-a_j, b_j) = C_1 - C_2 a_j + C_3 b_j
\tag{3.8}
$$

The value of C_1 is given directly by equation 3.6, and the values of C_2 and C_3

are obtained from equations 3.7 and 3.8 as

$$C_2 = \frac{-(b_j + b_k)\,w_i + b_j w_j + b_k w_k}{a_k b_j - a_j b_k} \qquad (3.9)$$

$$C_3 = \frac{-(a_j + a_k)\,w_i + a_j w_j + a_k w_k}{a_k b_j - a_j b_k} \qquad (3.10)$$

Using equations 3.3 and 3.4, these definitions may be expressed in matrix form as

$$\begin{bmatrix} C_2 \\ C_3 \end{bmatrix} = \frac{1}{2\Delta_m} B \begin{bmatrix} w_i \\ w_j \\ w_k \end{bmatrix} \qquad (3.11)$$

where B is a dimension matrix

$$B = \begin{bmatrix} b_i & b_j & b_k \\ a_i & a_j & a_k \end{bmatrix} \qquad (3.12)$$

Since the present analysis is formulated with velocity as the dependent variable, compatibility of strain rates as defined by equations 2.14 to 2.19 is automatically satisfied within each element. Compatibility, in the form of continuity of velocity, is also satisfied across the inter-element boundaries. The distribution of velocity along a side of an element between the values at the two nodal points is linear. Therefore, the distributions along the common boundaries of adjoining elements are identical. A more general view of inter-element compatibility is provided by the variational formulation, and is discussed in section 3.7.1.

3.1.2 Forces acting on the element   In order to proceed with the analysis it is necessary to determine the forces acting on individual elements. Since the present elements are of the simple CST type, the strain rates and hence the stresses are constant over each element. The nonzero viscous shear stresses are σ_{zx} and σ_{zy}: figure 3.3a shows these stresses acting on the rectangular prism enclosing the typical element shown in figure 3.2. The effects of the viscous stresses can be expressed in terms of equivalent forces acting at the mid-points of the sides of the element as shown in figure 3.3b. For example, consider the side joining the nodes i and j. The force in the z-direction at the mid-point of this side is due to a stress of $+\sigma_{zx}$ acting on an area of $(-b_k) \times 1$, together with a stress of $-\sigma_{zy}$ acting on an area of $a_k \times 1$, and is therefore equal to $-\sigma_{zx}b_k - \sigma_{zy}a_k$.

A further transformation allows the forces acting at the mid-points of the sides of the element to be replaced by an equivalent set acting at the nodes as shown in figure 3.3c. In order to maintain the same resultant force and moment about any point on a side of the element, the force at the mid-point must be

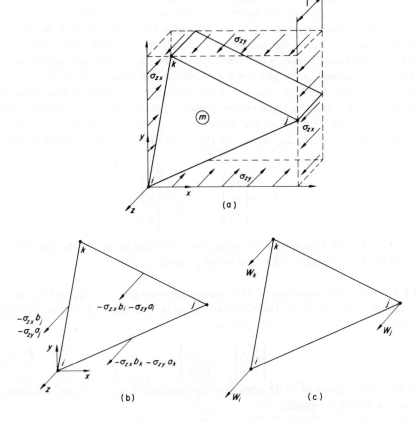

Figure 3.3 Equivalent systems of stresses and forces acting on an element: (a) stresses on the enclosing prism; (b) forces at the mid-points of the element sides; (c) forces at the nodes

replaced by two equal forces of half the magnitude at the relevant nodes. Thus, for example

$$W_i = -\tfrac{1}{2}(\sigma_{zx}b_k + \sigma_{zy}a_k) - \tfrac{1}{2}(\sigma_{zx}b_j + \sigma_{zy}a_j)$$

and using equations 3.3 this expression becomes

$$W_i = \tfrac{1}{2}(\sigma_{zx}b_i + \sigma_{zy}a_i)$$

Similar expressions may be obtained for W_j and W_k to give

$$\begin{bmatrix} W_i \\ W_j \\ W_k \end{bmatrix} = \tfrac{1}{2}\begin{bmatrix} b_i & a_i \\ b_j & a_j \\ b_k & a_k \end{bmatrix}\begin{bmatrix} \sigma_{zx} \\ \sigma_{zy} \end{bmatrix} = \tfrac{1}{2}B^{\mathrm{T}}\begin{bmatrix} \sigma_{zx} \\ \sigma_{zy} \end{bmatrix} \tag{3.13}$$

where B is the dimension matrix defined in equation 3.12 and the superscript T denotes a matrix transposition (interchange of rows and columns).

The only other forces acting on the element are those due to the downstream pressure gradient P_z, which according to equations 2.32 and 2.33 is constant over the solution domain. The pressure difference across the unit thickness of the typical element is P_z and the total pressure force on it is $-P_z \Delta_m$ in the z-direction. The negative sign arises because if P_z is positive the pressure increases with z and the force on the element is in the negative z-direction. The force acts at the centroid of the element, and is therefore equivalent to three equal forces of $-\frac{1}{3} P_z \Delta_m$ acting in the same direction at the nodal points. Hence

$$F_m = \begin{bmatrix} F_i \\ F_j \\ F_k \end{bmatrix} = -\frac{P_z \Delta_m}{3} \begin{bmatrix} 1 \\ 1 \\ 1 \end{bmatrix} \qquad (3.14)$$

where F_i, F_j and F_k are effectively the external forces applied in the z-direction to the mesh at nodes i, j and k, due to the presence of element m.

3.1.3 Constitutive equations   The viscous stress components σ_{zx} and σ_{zy} may be obtained in terms of the shear strain rates with the aid of constitutive equations 2.29 for a newtonian fluid as

$$\begin{bmatrix} \sigma_{zx} \\ \sigma_{zy} \end{bmatrix} = \mu \begin{bmatrix} e_{zx} \\ e_{zy} \end{bmatrix} \qquad (3.15)$$

From the definitions of strain rates given by equations 2.4 and 2.5, and the shape function shown in equation 3.5

$$\begin{bmatrix} e_{zx} \\ e_{zy} \end{bmatrix} = \begin{bmatrix} \dfrac{\partial w}{\partial x} \\[2mm] \dfrac{\partial w}{\partial y} \end{bmatrix} = \begin{bmatrix} C_2 \\ C_3 \end{bmatrix} \qquad (3.16)$$

Therefore, using equation 3.11

$$\begin{bmatrix} \sigma_{zx} \\ \sigma_{zy} \end{bmatrix} = \frac{\mu}{2\Delta_m} B \begin{bmatrix} w_i \\ w_j \\ w_k \end{bmatrix} \qquad (3.17)$$

Substitution of this relationship into equation 3.13 allows the viscous forces acting on the element at its nodes to be expressed in terms of the corresponding velocities

$$\begin{bmatrix} W_i \\ W_j \\ W_k \end{bmatrix} = \frac{\mu}{4\Delta_m} B^{\mathrm{T}} B \begin{bmatrix} w_i \\ w_j \\ w_k \end{bmatrix} \qquad (3.18)$$

This result may be expressed in the general notation introduced in section 1.2.2 as

$$W_m = k_m \delta_m \tag{3.19}$$

where δ_m is the element velocity (displacement) vector and k_m is the element (viscous) stiffness matrix. This stiffness matrix is given by

$$k_m = \frac{\mu}{4\Delta_m} B^T B \tag{3.20}$$

$$= \frac{\mu}{4\Delta_m} \begin{bmatrix} a_i^2 + b_i^2 & a_i a_j + b_i b_j & a_k a_i + b_k b_i \\ a_i a_j + b_i b_j & a_j^2 + b_j^2 & a_j a_k + b_j b_k \\ a_k a_i + b_k b_i & a_j a_k + b_j b_k & a_k^2 + b_k^2 \end{bmatrix} \tag{3.21}$$

and is symmetric.

3.2   Assembly of the Overall Stiffness Matrix

The actual viscous stresses and pressure differences acting on individual elements have been replaced by the equivalent forces acting at the nodes of the mesh. The conditions required for equilibrium can be expressed in general as

$$\begin{pmatrix} \text{externally applied} \\ \text{forces at the nodes} \end{pmatrix} = \Sigma \begin{pmatrix} \text{forces on the elements} \\ \text{at these nodes} \end{pmatrix}$$

and for the present problem

$$\Sigma \begin{pmatrix} \text{pressure forces} \\ \text{acting at the nodes} \end{pmatrix} = \Sigma \begin{pmatrix} \text{viscous forces on the} \\ \text{elements at these nodes} \end{pmatrix}$$

For example, for equilibrium of forces acting at node i

$$\Sigma F_i^{(m)} = \Sigma W_i^{(m)} \tag{3.22}$$

where $F_i^{(m)}$ is the pressure force at the node due to the presence of element m, and $W_i^{(m)}$ is the viscous force on the same element, defined according to equations 3.14 and 3.18. The summations indicated in equation 3.22 are performed for elements which have the point i as a node.

The set of equations for equilibrium of all the nodes can be expressed in the general form

$$K\delta = F \tag{3.23}$$

where

$$K\delta = \Sigma W_m = \Sigma k_m \delta_m \tag{3.24}$$

and

$$F = \Sigma F_m \tag{3.25}$$

The vector δ contains the velocities at the nodal points: if there are n nodes numbered from 1 to n, δ contains w_1, w_2, w_3, ..., w_n in this order. The vector F contains the total pressure forces acting at the nodes, in the same order. The square matrix K is the overall (viscous) stiffness matrix, and is sparse because only a relatively small number of elements have any particular point as a node.

Let K_{pq} and k_{rs} be typical coefficients of the overall and element stiffness matrices respectively, where p and q lie in the range 1 to n, while r and s lie in the range 1 to 3. The subscripts p and r are matrix row numbers, while q and s are column numbers. Now k_{rs} can be interpreted as the viscous force that must be applied to the typical element at the node corresponding to the rth coefficient of the force vector shown in equation 3.18 to cause a unit velocity at the node corresponding to the sth coefficient of the element velocity vector. Similarly, K_{pq} can be interpreted as the force that must be applied to the overall system at node p to produce a unit velocity at node q. The process of assembling the overall stiffness coefficients takes the form of

$$K_{pq} = \Sigma k_{rs} \qquad (3.26)$$

where the row and column numbers are equivalent: $p = i$, j or k according to whether $r = 1$, 2 or 3, and similarly for q according to the value of s. Assembly is complete when the nine coefficients of each and every element stiffness matrix have been added to the relevant overall stiffness coefficients.

The overall pressure force vector is assembled as indicated by equations 3.25 and 3.22.

$$F_p = -\Sigma \tfrac{1}{3} P_z \Delta_m \qquad (3.27)$$

where the summation is performed for those elements which have the point p as a node.

The assembly process can be illustrated with the aid of the very simple two-element mesh shown in figure 3.4. The four nodal points are numbered $a, b,$

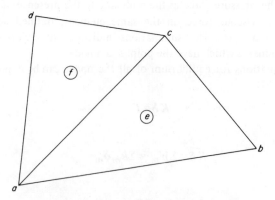

Figure 3.4   A simple two-element mesh

c and d and the elements e and f. For element e

$$\begin{bmatrix} W_a \\ W_b \\ W_c \end{bmatrix}_e = \begin{bmatrix} k_{11} & k_{12} & k_{13} \\ k_{21} & k_{22} & k_{23} \\ k_{31} & k_{32} & k_{33} \end{bmatrix}_e \begin{bmatrix} w_a \\ w_b \\ w_c \end{bmatrix}_e \qquad (3.28)$$

where the stiffness coefficients are derived from the material viscosity and element geometry according to equation 3.21. Similarly, for element f

$$\begin{bmatrix} W_a \\ W_c \\ W_d \end{bmatrix}_f = \begin{bmatrix} k_{11} & k_{12} & k_{13} \\ k_{21} & k_{22} & k_{23} \\ k_{31} & k_{32} & k_{33} \end{bmatrix}_f \begin{bmatrix} w_a \\ w_c \\ w_d \end{bmatrix}_f \qquad (3.29)$$

Note that in each case the nodal point numbers are taken in an anticlockwise order round the element: the two elements need not have the same local origin. The pressure forces acting at the nodes due to the pressure gradient acting on the two elements are

$$\begin{bmatrix} F_a \\ F_b \\ F_c \end{bmatrix}_e = -\frac{P_z \Delta_e}{3} \begin{bmatrix} 1 \\ 1 \\ 1 \end{bmatrix}, \quad \begin{bmatrix} F_a \\ F_c \\ F_d \end{bmatrix}_f = -\frac{P_z \Delta_f}{3} \begin{bmatrix} 1 \\ 1 \\ 1 \end{bmatrix} \qquad (3.30)$$

For equilibrium of forces acting on the four nodes, equation 3.22 can be used to give

$$\begin{bmatrix} (k_{11}^{(e)} + k_{11}^{(f)}) & k_{12}^{(e)} & (k_{13}^{(e)} + k_{12}^{(f)}) & k_{13}^{(f)} \\ k_{21}^{(e)} & k_{22}^{(e)} & k_{23}^{(e)} & 0 \\ (k_{31}^{(e)} + k_{21}^{(f)}) & k_{32}^{(e)} & (k_{33}^{(e)} + k_{22}^{(f)}) & k_{23}^{(f)} \\ k_{31}^{(f)} & 0 & k_{32}^{(f)} & k_{33}^{(f)} \end{bmatrix} \begin{bmatrix} w_a \\ w_b \\ w_c \\ w_d \end{bmatrix} = -\frac{P_z}{3} \begin{bmatrix} \Delta_e + \Delta_f \\ \Delta_e \\ \Delta_e + \Delta_f \\ \Delta_f \end{bmatrix} \qquad (3.31)$$

where the superscripts on the stiffness coefficients refer to the numbers of the elements from which they are derived. The overall stiffness and pressure force coefficients could equally well have been obtained with the aid of equations 3.26 and 3.27. Another relatively simple example of the assembly process is given in the next section.

Equation 3.21 shows that the stiffness matrix for any element is symmetric, that is, $k_{rs} = k_{sr}$. Consequently the overall stiffness matrix displayed in equation 3.31 is also symmetric, as are overall stiffness matrices in general for materials with constitutive equations of the simple linear type discussed in section 2.1.6.

3.3 Comparison with the Finite Difference Approach

It is interesting to compare in detail the equations established by the present finite element analysis with those derived from the finite difference approach described in section 2.4.1. Equation 2.92 provides a finite difference approximation for the governing harmonic equation 2.87 at the point labelled O in figure 2.7, in terms of the values of the dependent variable ψ at the surrounding points W, E, S and N. In the present problem of downstream viscous flow, equation 2.31 is the particular form of the general harmonic equation 2.87, and $\psi \equiv w, \phi_1 \equiv P_z/\mu$. Equation 2.92 therefore becomes

$$\frac{w_W - 2w_O + w_E}{h_x^2} + \frac{w_S - 2w_O + w_N}{h_y^2} = \frac{P_z}{\mu} \tag{3.32}$$

In order to obtain a comparable result from the finite element analysis it is necessary to choose a mesh which connects the point O only to the same four points W, E, S and N, as shown in figure 3.5. Let the four elements which have point O as a node be numbered e, f, g and h as shown. For element e, if nodes O E and N are taken to be equivalent to i, j and k respectively of the typical element shown in figure 3.2, then according to equations 3.1

$$\begin{aligned}
a_i &= -h_x, & b_i &= -h_y \\
a_j &= 0, & b_j &= h_y \\
a_k &= h_x, & b_k &= 0
\end{aligned} \tag{3.33}$$

From equations 3.18 and 3.21 the viscous forces acting on the element at its nodes are given by

$$\begin{bmatrix} W_O \\ W_E \\ W_N \end{bmatrix}_e = \frac{\mu}{2h_x h_y} \begin{bmatrix} h_x^2 + h_y^2 & -h_y^2 & -h_x^2 \\ -h_y^2 & h_y^2 & 0 \\ -h_x^2 & 0 & h_x^2 \end{bmatrix}_e \begin{bmatrix} w_O \\ w_E \\ w_N \end{bmatrix}_e \tag{3.34}$$

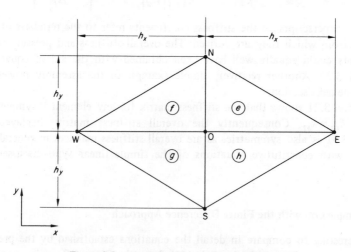

Figure 3.5　A mesh of four right-angled elements

and the pressure forces are given by equation 3.14 as

$$\begin{bmatrix} F_O \\ F_E \\ F_N \end{bmatrix}_e = -\frac{P_z h_x h_y}{6} \begin{bmatrix} 1 \\ 1 \\ 1 \end{bmatrix} \tag{3.35}$$

Identical sets of pressure force components may be shown to act at the nodes of the other three elements. The stiffness matrix for element g is identical with the one displayed in equation 3.34, provided point O is again taken to be the origin (equivalent to i for the typical element). On the other hand, if O is the origin for element f

$$\begin{aligned} a_i &= -h_x, & b_i &= h_y \\ a_j &= h_x, & b_j &= 0 \\ a_k &= 0, & b_k &= -h_y \end{aligned} \tag{3.36}$$

and the viscous forces are given by

$$\begin{bmatrix} W_O \\ W_N \\ W_W \end{bmatrix}_f = \frac{\mu}{2h_x h_y} \begin{bmatrix} h_x^2 + h_y^2 & -h_x^2 & -h_y^2 \\ -h_x^2 & h_x^2 & 0 \\ -h_y^2 & 0 & h_y^2 \end{bmatrix}_f \begin{bmatrix} w_O \\ w_N \\ w_W \end{bmatrix}_f \tag{3.37}$$

The same stiffness matrix is appropriate for element h if O is again used as the origin. For equilibrium of forces acting on the five nodes, equation 3.22 can be used to give

$$\frac{\mu}{2h_x h_y} \begin{bmatrix} 4(h_x^2 + h_y^2) & -2h_y^2 & -2h_x^2 & -2h_y^2 & -2h_x^2 \\ -2h_y^2 & 2h_y^2 & 0 & 0 & 0 \\ -2h_x^2 & 0 & 2h_x^2 & 0 & 0 \\ -2h_y^2 & 0 & 0 & 2h_y^2 & 0 \\ -2h_x^2 & 0 & 0 & 0 & 2h_x^2 \end{bmatrix} \begin{bmatrix} w_O \\ w_E \\ w_N \\ w_W \\ w_S \end{bmatrix} = -\frac{P_z h_x h_y}{6} \begin{bmatrix} 4 \\ 2 \\ 2 \\ 2 \\ 2 \end{bmatrix} \tag{3.38}$$

The first of these equations, which expresses the equilibrium condition for the forces on point O, can be rewritten as

$$\frac{w_W - 2w_O + w_E}{h_x^2} + \frac{w_S - 2w_O + w_N}{h_y^2} = \frac{2P_z}{3\mu} \tag{3.39}$$

the left-hand side of which is identical with the left-hand side of the finite difference equation 3.32. The fact that the right-hand sides are not also identical is due to the use of an 'irregular' mesh of triangular elements, which is necessary to obtain the required connections between point O and the four surrounding points. Figure 3.6 shows a larger portion of the complete mesh: it is irregular in

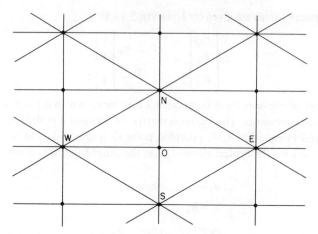

Figure 3.6   Part of a mesh containing the four-element mesh shown in figure 3.5

the sense that, whereas point O serves as a node for only four elements, points such as W, E, S and N are associated with eight elements. Consequently the sum of the pressure forces at such nodes is twice that for point O. For example, the equation of equilibrium for the forces acting on point N expressed in a form equivalent to equation 3.39 would have $4P_z/3\mu$ on its right-hand side. Therefore, the average right-hand side in the equilibrium equations for all the points in the mesh would be P_z/μ, in agreement with the finite difference form.

In view of the close similarity between the equations obtained from the present finite element and finite difference methods, it is to be expected that their orders of accuracy are similar. Truncation errors involved in the finite difference method are discussed in section 2.4.1, and those for the finite element method are examined in section 3.7.2.

3.4   Variational Formulation

The use of a variational principle provides an alternative method of formulating a finite element analysis. This approach, although mathematically more sophisticated than the direct equilibrium method, offers a number of advantages. It shows how a particular analysis and computer program can be generalised to solve any problem of the same mathematical type. Its use is necessary if elements more complex than the CST type are employed, as discussed in section 8.2. Indeed, in some books there is a tendency to treat variational formulation as an essential feature of finite element methods. Finally, it helps to clarify both the method of applying boundary conditions (section 3.5) and the conditions necessary for convergence of finite element methods (section 3.7). Nevertheless, this section could well be omitted on a first reading of this chapter.

3.4.1 The variational principle The general variational approach to the solution of a continuum mechanics problem is to seek a stationary value (often a minimum) for a quantity χ which is defined by an appropriate integration of the unknowns over the solution domain. Such a quantity χ (which should not be confused with the stress functions introduced in chapter 2) is often referred to as a 'functional'. When such a principle is used in a finite element analysis, the variation of χ is carried out with respect to the values of the unknowns at the nodes of the mesh.

Sometimes the relevant functional can be readily obtained directly from familiar physical principles. For example, the solutions to structural and elastic continuum problems can be obtained by minimising the total potential energy (see, for example, Zienkiewicz (1971)). In general, however, the choice of functional is less straightforward (see Schechter (1967)), unless explicit use is made of the governing differential equation. For present purposes, solutions to the general two-dimensional harmonic equation 2.87 are required. That is

$$\nabla^2 \psi = \phi_1(x, y) \tag{3.40}$$

where ϕ_1 is a known function of position in the solution domain in the $x-y$ plane.

In order to find the required functional it is convenient to let

$$\lambda = \nabla^2 \psi - \phi_1(x, y) = 0 \tag{3.41}$$

Now ψ is a continuous function of position which in general can only be defined exactly in terms of an infinite number of parameters, such as values of the function at particular points in the solution domain. The object of the present analysis is to provide a means of determining an approximate form of ψ in terms of a finite number of parameters. Let η be a typical such parameter, and multiply every term in equation 3.41 by the derivative of ψ with respect to η to give

$$\lambda \frac{\partial \psi}{\partial \eta} = \frac{\partial \psi}{\partial \eta} \frac{\partial^2 \psi}{\partial x^2} + \frac{\partial \psi}{\partial \eta} \frac{\partial^2 \psi}{\partial y^2} - \frac{\partial \psi}{\partial \eta} \phi_1(x, y) = 0 \tag{3.42}$$

Since λ is zero everywhere within the solution domain

$$\int \int \int \lambda \frac{\partial \psi}{\partial \eta} \, d(\text{volume}) = 0 \tag{3.43}$$

where the integration is performed over the entire domain, which in general is three-dimensional. For the typical two-dimensional domain shown in figure 3.7a, equation 3.43 becomes

$$\int \int \lambda \frac{\partial \psi}{\partial \eta} \, dx \, dy = 0 \tag{3.44}$$

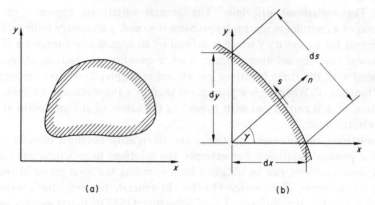

Figure 3.7　A typical two-dimensional solution domain: (a) the entire domain; (b) part of the domain boundary

Therefore, using equation 3.42

$$\iint \left[\frac{\partial \psi}{\partial \eta} \frac{\partial^2 \psi}{\partial x^2} + \frac{\partial \psi}{\partial \eta} \frac{\partial^2 \psi}{\partial y^2} - \frac{\partial \psi}{\partial \eta} \phi_1(x, y) \right] dx \, dy = 0 \qquad (3.45)$$

Now

$$\frac{\partial \psi}{\partial \eta} \frac{\partial^2 \psi}{\partial x^2} = \frac{\partial}{\partial x} \left(\frac{\partial \psi}{\partial \eta} \frac{\partial \psi}{\partial x} \right) - \frac{1}{2} \frac{\partial}{\partial \eta} \left[\left(\frac{\partial \psi}{\partial x} \right)^2 \right]$$

and

$$\frac{\partial \psi}{\partial \eta} \frac{\partial^2 \psi}{\partial y^2} = \frac{\partial}{\partial y} \left(\frac{\partial \psi}{\partial \eta} \frac{\partial \psi}{\partial y} \right) - \frac{1}{2} \frac{\partial}{\partial \eta} \left[\left(\frac{\partial \psi}{\partial y} \right)^2 \right]$$

and since $\phi_1(x, y)$ is independent of η, equation 3.45 becomes

$$\iint \frac{\partial}{\partial \eta} \left[\frac{1}{2} \left(\frac{\partial \psi}{\partial x} \right)^2 + \frac{1}{2} \left(\frac{\partial \psi}{\partial y} \right)^2 + \psi \phi_1(x, y) \right] dx \, dy - I = 0 \qquad (3.46)$$

where

$$I = \iint \left[\frac{\partial}{\partial x} \left(\frac{\partial \psi}{\partial \eta} \frac{\partial \psi}{\partial x} \right) + \frac{\partial}{\partial y} \left(\frac{\partial \psi}{\partial \eta} \frac{\partial \psi}{\partial y} \right) \right] dx \, dy$$

Applying Green's theorem

$$I = \oint \left(\frac{\partial \psi}{\partial \eta} \frac{\partial \psi}{\partial x} \, dy - \frac{\partial \psi}{\partial \eta} \frac{\partial \psi}{\partial y} \, dx \right)$$

where the line integration is performed in the anticlockwise direction around the boundary of the solution domain. If n is the direction of the outward normal to the boundary at a particular point as shown in figure 3.7b

$$\frac{\partial \psi}{\partial n} = \frac{\partial \psi}{\partial x} \cos \gamma + \frac{\partial \psi}{\partial y} \sin \gamma$$

where γ is the angle between the normal and the x-axis. Now

$$\cos \gamma = \frac{dy}{ds}, \quad \sin \gamma = -\frac{dx}{ds}$$

where s is distance along the boundary measured in the anticlockwise direction, and the negative sign is due to the fact that, for positive $\sin \gamma$, x decreases as s increases. Hence

$$\frac{\partial \psi}{\partial n} ds = \frac{\partial \psi}{\partial x} dy - \frac{\partial \psi}{\partial y} dx$$

and

$$I = \oint \frac{\partial \psi}{\partial \eta} \frac{\partial \psi}{\partial n} ds \tag{3.47}$$

Now if the conditions on the boundary are such that either the value of ψ is prescribed and therefore independent of η, or the value of its first derivative normal to the boundary is zero, then $I = 0$. The more general case when this is not so is considered in section 3.5.

According to equation 3.46, the solution of the governing differential equation 3.40 is obtained when the value of the functional

$$\chi = \int \int \left[\frac{1}{2} \left(\frac{\partial \psi}{\partial x} \right)^2 + \frac{1}{2} \left(\frac{\partial \psi}{\partial y} \right)^2 + \psi \, \phi_1 (x, y) \right] dx \, dy \tag{3.48}$$

is stationary with respect to the parameter η, provided $I = 0$. For example, a particular mathematical function of position involving a finite number of undetermined parameters, such as a polynomial with undetermined coefficients, might be assumed for ψ. The best fit of this approximation to the true distribution could be obtained by making the value of χ stationary with respect to the function parameters. Finite element methods can be regarded as being of this type. Instead of using a single and necessarily complicated function to cover the whole solution domain, a series of relatively simple functions applicable to small subregions is employed. The overall distribution of ψ is thereby defined in terms of values at points of interconnection between the subregions. The required solution is obtained when the value of χ is stationary with respect to each and every one of these point values, ψ_i; that is, when

$$\frac{\partial \chi}{\partial \psi_i} = 0 \tag{3.49}$$

3.4.2 Application to the downstream viscous flow problem   Finite element analyses can be formulated by the variational approach by first assuming that the total value of the functional χ may be obtained by summing the results of the relevant integrations performed over all the individual elements

$$\chi = \Sigma \chi^{(m)} \tag{3.50}$$

The validity of this summation is only assured if the convergence conditions discussed in section 3.7.1 (equivalent to the inter-element compatibility conditions discussed in section 3.1.1) are satisfied.

For the downstream viscous flow problem, the general harmonic equation 3.40 takes the particular form of equation 2.31, and

$$\chi^{(m)} = \int\int \left[\frac{1}{2}\left(\frac{\partial w}{\partial x}\right)^2 + \frac{1}{2}\left(\frac{\partial w}{\partial y}\right)^2 + \frac{wP_z}{\mu} \right] dx\ dy \tag{3.51}$$

the integration being performed over the area of the particular element, numbered m. If the element is of the CST type described in section 3.1.1 then

$$\chi^{(m)} = \int\int \left[\frac{1}{2}(C_2^2 + C_3^2) + \frac{wP_z}{\mu} \right] dx\ dy \tag{3.52}$$

Since the velocity w is the only variable in the integrand and is a linear function of x and y

$$\chi^{(m)} = \Delta_m \left[\frac{1}{2}(C_2^2 + C_3^2) + \frac{\bar{w}P_z}{\mu} \right] \tag{3.53}$$

where the mean velocity $\bar{w}$ is the average of the values at the nodes of the element

$$\bar{w} = \frac{1}{3}(w_i + w_j + w_k) \tag{3.54}$$

According to the variational principle the required solution is obtained when the value of χ is stationary with respect to the nodal point velocities; that is, when

$$\frac{\partial \chi}{\partial w_i} = \Sigma \frac{\partial \chi^{(m)}}{\partial w_i} = 0 \tag{3.55}$$

For a particular point i, the summation need only be performed for elements which have i as a node. Using equations 3.53 and 3.54

$$\frac{\partial \chi^{(m)}}{\partial w_i} = \Delta_m \left(C_2 \frac{\partial C_2}{\partial w_i} + C_3 \frac{\partial C_3}{\partial w_i} + \frac{P_z}{3\mu} \right) \tag{3.56}$$

and the derivatives of $\chi^{(m)}$ with respect to the three nodal point velocities of element m may be expressed in matrix form as

$$
\begin{bmatrix}
\dfrac{\partial \chi^{(m)}}{\partial w_i} \\[2ex]
\dfrac{\partial \chi^{(m)}}{\partial w_j} \\[2ex]
\dfrac{\partial \chi^{(m)}}{\partial w_k}
\end{bmatrix}
= \Delta_m
\begin{bmatrix}
\dfrac{\partial C_2}{\partial w_i} & \dfrac{\partial C_3}{\partial w_i} \\[2ex]
\dfrac{\partial C_2}{\partial w_j} & \dfrac{\partial C_3}{\partial w_j} \\[2ex]
\dfrac{\partial C_2}{\partial w_k} & \dfrac{\partial C_3}{\partial w_k}
\end{bmatrix}
\begin{bmatrix}
C_2 \\[2ex]
C_3
\end{bmatrix}
+ \frac{P_z \Delta_m}{3\mu}
\begin{bmatrix}
1 \\[1ex]
1 \\[1ex]
1
\end{bmatrix}
\tag{3.57}
$$

Now, the constants C_2 and C_3 are given in terms of the nodal point velocities by equation 3.11, which also yields the result

$$
\begin{bmatrix}
\dfrac{\partial C_2}{\partial w_i} & \dfrac{\partial C_3}{\partial w_i} \\[2ex]
\dfrac{\partial C_2}{\partial w_j} & \dfrac{\partial C_3}{\partial w_j} \\[2ex]
\dfrac{\partial C_2}{\partial w_k} & \dfrac{\partial C_3}{\partial w_k}
\end{bmatrix}
= \frac{1}{2\Delta_m} B^{\mathrm{T}}
\tag{3.58}
$$

where B is the dimension matrix defined in equation 3.12. Hence

$$
\begin{bmatrix}
\dfrac{\partial \chi^{(m)}}{\partial w_i} \\[2ex]
\dfrac{\partial \chi^{(m)}}{\partial w_j} \\[2ex]
\dfrac{\partial \chi^{(m)}}{\partial w_k}
\end{bmatrix}
= \frac{1}{4\Delta_m} B^{\mathrm{T}} B
\begin{bmatrix}
w_i \\[2ex]
w_j \\[2ex]
w_k
\end{bmatrix}
+ \frac{P_z \Delta_m}{3\mu}
\begin{bmatrix}
1 \\[1ex]
1 \\[1ex]
1
\end{bmatrix}
\tag{3.59}
$$

Equation 3.55 is therefore identical with the direct equilibrium condition given by equation 3.22. This result is to be expected because the differential equation on which the variational formulation is based is derived from an equation of equilibrium. For present purposes, the main practical use of the variational formulation is to show how any problem of the harmonic type may be solved by the same computer program.

3.5 Boundary Conditions

The overall set of linear equations 3.23 has been derived by means of both the direct equilibrium and variational formulations, without introducing the boundary conditions for a particular problem. This is in contrast to the finite difference approach described in section 2.4.1 in which special equations have to be derived for all points on the boundary of the solution domain. Types of boundary conditions are discussed in section 2.2.9: the value of either the dependent variable or its first derivative normal to the boundary is often prescribed. A more general condition is provided by equation 2.84. The 'restraint' conditions referred to in section 1.2.2 for rigid-jointed structures are in effect boundary conditions.

Prescribed velocity boundary conditions may be applied by modifying equations 3.23. If the velocity at boundary nodal point p, say, is to take the value α, the pth equation is modified to be equivalent to $\delta_p = \alpha$. There are two ways in which the coefficients of K and F can be modified to achieve this effect. In the first method the required form is obtained explicitly by setting

$$K_{pp}^* = 1, \quad F_p^* = \alpha, \quad K_{pq}^* = 0 \quad \text{for all } q, q \neq p \tag{3.60}$$

where the asterisks indicate modified coefficients. An illustration of this technique is provided by the structural analysis case study described in section 1.3. Since the value of δ_p is prescribed, it can be removed as a variable from the remaining equations by setting

$$K_{qp}^* = 0, \quad F_q^* = F_q - \alpha K_{qp}, \quad \text{for all } q, q \neq p \tag{3.61}$$

which returns the stiffness matrix to a symmetric form. In the second method the result $\delta_p = \alpha$ is achieved indirectly but more simply by making the modifications

$$K_{pp}^* = MK_{pp}, \quad F_p^* = \alpha K_{pp}^* \tag{3.62}$$

where M is a very large number, say of the order of 10^{10}. The effect is to make the nondiagonal coefficients in the pth row of the stiffness matrix negligible compared with the diagonal one, leaving the pth equation as

$$K_{pp}^* \delta_p = \alpha K_{pp}^* \tag{3.63}$$

Zero normal derivative boundary conditions of the form $\partial \delta / \partial n = 0$ are even easier to apply. In the context of the downstream viscous flow problem, this condition implies that there is no shear stress acting on the boundary. In other words, the presence of the boundary has no effect on the velocity distribution: boundary nodal points are treated as internal nodes and equations 3.23 are unchanged.

The general boundary condition displayed in equation 2.84 can be accommodated with the aid of the variational formulation. Equation 3.48 is derived on the assumption that the integral I defined in equation 3.47 takes the value zero. If on part of the boundary, S, the dependent variable ψ is governed by this general condition then

$$I = -\int_S \frac{\partial \psi}{\partial \eta} \left(\frac{a_2}{a_1} \psi + \frac{a_3}{a_1} \right) ds = -\int_S \frac{\partial}{\partial \eta} \left(\frac{a_2}{a_1} \frac{\psi^2}{2} + \frac{a_3}{a_1} \psi \right) ds \tag{3.64}$$

and hence to χ must be added the term

$$\int_S \left(\frac{a_2}{a_1} \frac{\psi^2}{2} + \frac{a_3}{a_1} \psi \right) \mathrm{d}s \tag{3.65}$$

which can take the form of additional stiffnesses for elements at the boundary. Incidentally, this result serves to confirm that the zero normal derivative boundary condition ($a_2 = a_3 = 0$) requires no modification of the overall set of equations. The case where the value of ψ is prescribed on the boundary ($a_1 = 0$) requires the special treatment already described.

3.6   Solution of the Linear Equations

General methods and computer programs for solving sets of simultaneous linear equations are described by Fenner (1974). The most commonly used techniques are the direct gaussian elimination and iterative Gauss–Seidel methods, the basic forms of which are described in appendixes A and B respectively. The elimination method is used in the structural analysis case study described in section 1.3. The purpose of this section is to review the application of these methods to the solution of finite element equations such as equations 3.23.

3.6.1 Gaussian elimination   A serious disadvantage of the basic gaussian elimination method described in appendix A is that it requires a large amount of computer storage capacity. In finite element applications the whole of the overall stiffness matrix must be stored, despite the fact that it is only sparsely populated with nonzero coefficients. For example, in problems of the harmonic type considered in this chapter, the number of nonzero coefficients in the ith row or column of the overall stiffness matrix is equal to the number of other nodes to which the point i is directly connected, plus one for the self-stiffness coefficient. In the case of the mesh shown in figure 3.6, for example, there are five and nine nonzero stiffness coefficients associated with the points O and N respectively.

Referring to appendix A, the kth equation need only be used to eliminate the kth unknown from the subsequent equations in which the coefficient of this unknown, $a_{ik}^{(k)}$, is nonzero: either because it was nonzero in the original equations or because it has been made so by previous eliminations. If the equations are arranged so that the nonzero coefficients are confined to a relatively narrow band parallel to the diagonal of the matrix, eliminations need only be performed up to the extent of this band. Only the coefficients within the band need be stored, since the rest remain zero throughout the solution process. Consider the example of a tridiagonal matrix, in which all coefficients other than those on the diagonal or immediately adjacent to it are zero.

$$A = \begin{bmatrix} 1 & 1 & 0 & 0 \\ 1 & 2 & 1 & 0 \\ 0 & 1 & 2 & 1 \\ 0 & 0 & 1 & 0 \end{bmatrix}$$

(3.66)

The result of applying the elimination process is

$$A^* = \begin{bmatrix} 1 & 1 & 0 & 0 \\ 0 & 1 & 1 & 0 \\ 0 & 0 & 1 & 1 \\ 0 & 0 & 0 & -1 \end{bmatrix}$$

(3.67)

Coefficients outside the original bandwidth of three remain as zero, whereas zeros within the band in general do not.

A banded square stiffness matrix can be stored as a rectangular one with a width equal to the bandwidth: the particular case of a tridiagonal matrix is considered in detail by Fenner (1974). If the matrix is also symmetric a further saving can be made by storing only half of the band of nonzero coefficients. Elimination methods and computer programs for solving banded sparse sets of equations are discussed by Zienkiewicz (1971) and Nath (1974).

It should be noted that it is the bandwidth rather than the number of nonzero coefficients per row that determines the storage requirement for a finite element stiffness matrix. If, for example, the mesh used involves a total of n nodal points arranged in $\sqrt{n}$ rows of $\sqrt{n}$ points as shown in figure 4.2, then a typical node is connected not only to adjacent points in the same row, but also to points in the adjacent rows. For large values of n the bandwidth of the stiffness matrix is therefore of the order

$$b \approx 2\sqrt{n}$$

(3.68)

3.6.2 The Gauss–Seidel method   The basic Gauss–Seidel method described in appendix B can be readily adapted to minimise the amount of computer storage required for sparse sets of equations. In finite element applications the square overall stiffness matrix can be stored as a rectangular matrix with the number of columns equal to the maximum number of nonzero coefficients per row. For example, the stiffness matrix

$$K = \begin{bmatrix} 3.0 & 0 & 2.0 & 0 & 0 \\ 0 & 5.0 & 1.0 & 0 & 4.0 \\ 2.0 & 1.0 & 3.0 & 0 & 0 \\ 0 & 0 & 0 & 2.0 & 2.0 \\ 0 & 4.0 & 0 & 2.0 & 6.0 \end{bmatrix}$$

(3.69)

can be stored as follows

$$\tilde{K} = \begin{bmatrix} 3.0 & 2.0 & 0 \\ 5.0 & 4.0 & 1.0 \\ 3.0 & 2.0 & 1.0 \\ 2.0 & 2.0 & 0 \\ 6.0 & 2.0 & 4.0 \end{bmatrix}, \quad M = \begin{bmatrix} 1 & 3 & 0 \\ 2 & 5 & 3 \\ 3 & 1 & 2 \\ 4 & 5 & 0 \\ 5 & 4 & 2 \end{bmatrix}, \quad L = \begin{bmatrix} 2 \\ 3 \\ 3 \\ 2 \\ 3 \end{bmatrix} \quad (3.70)$$

The matrix $\tilde{K}$ is the rectangular form of K, the coefficients of M store the original column numbers in K of the corresponding coefficients in $\tilde{K}$, and the coefficients of vector L store the numbers of nonzero coefficients in the corresponding rows of either K or $\tilde{K}$. The zero coefficients in M have no significance since they are not used.

In the present context, the coefficient in the ith row of L defines how many nodes are adjacent and directly connected to point i, including the point itself, and the coefficients in the same row of M define the numbers assigned to these nodes. The order of arrangement of coefficients along rows of $\tilde{K}$ could be arbitrary. For present purposes, however, it is convenient to store the diagonal of K as the first column of $\tilde{K}$. Thereafter the order is arbitrary for the nonzero coefficients, which are stored before any zeros are accommodated. In this simple example, the use of $\tilde{K}$, M and L in place of K offers no saving in the amount of storage required. With large stiffness matrices containing much smaller proportions of nonzero coefficients, the saving can be very considerable.

In the basic Gauss–Seidel method, changes in the unknowns between successive cycles of iteration are computed with the aid of equation B.6. Applied to equations 3.23

$$\Delta\delta_i = \frac{1}{K_{ii}} \left(F_i - \sum_{j=1}^{n} K_{ij}\delta_j \right) \quad (3.71)$$

and if K is stored in the rectangular form defined by $\tilde{K}$, M and L

$$\Delta\delta_i = \frac{1}{\tilde{K}_{i1}} \left(F_i - \sum_{l=1}^{L_i} \tilde{K}_{il}\delta_j \right) \quad (3.72)$$

where

$$j = M_{il}$$

The main disadvantage of the Gauss–Seidel method is that it may require a large amount of computing time to achieve convergence, and may indeed fail to converge at all. The sufficient condition for convergence is stated in appendix B as that of diagonal dominance of the coefficient matrix (overall stiffness matrix) as defined by equation B.3. According to equation 3.21, the coefficients on the diagonals of the present element stiffness matrices all have the same sign (positive). Diagonal dominance of each and every element stiffness matrix is

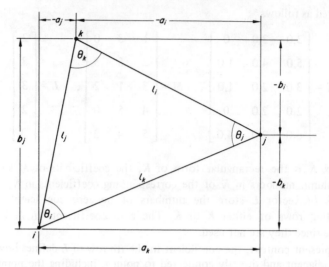

Figure 3.8   Angles and lengths of sides of a typical triangular element

therefore sufficient to ensure diagonal dominance of the overall stiffness matrix (see, for example, the matrices displayed in equations 3.31 and 3.38).

The diagonal dominance condition for an element stiffness matrix may be examined with the aid of figure 3.8, in which the lengths of the sides of the typical element are defined as l_i, l_j and l_k, and the angles at the corners as θ_i, θ_j and θ_k. The first row of stiffness coefficients shown in equation 3.21 is dominated by the one on the diagonal of the matrix if

$$a_i^2 + b_i^2 \geqslant |a_i a_j + b_i b_j| + |a_k a_i + b_k b_i| \tag{3.73}$$

Now

$$a_i^2 + b_i^2 = l_i^2$$

and

$$2a_i a_j = (a_i + a_j)^2 - a_i^2 - a_j^2$$
$$= a_k^2 - a_i^2 - a_j^2$$

Similarly

$$2b_i b_j = b_k^2 - b_i^2 - b_j^2$$
$$2a_k a_i = a_j^2 - a_k^2 - a_i^2$$
$$2b_k b_i = b_j^2 - b_k^2 - b_i^2$$

and equation 3.73 becomes

$$l_i^2 \geqslant \tfrac{1}{2}|l_k^2 - l_i^2 - l_j^2| + \tfrac{1}{2}|l_j^2 - l_k^2 - l_i^2| \tag{3.74}$$

Now, for $\theta_k < 90°$, $l_k^2 < l_i^2 + l_j^2$ and similarly for $\theta_j < 90°$, $l_j^2 < l_k^2 + l_i^2$; equation

3.74 becomes

$$l_i^2 \geq \frac{1}{2}(l_i^2 + l_j^2 - l_k^2) + \frac{1}{2}(l_k^2 + l_i^2 - l_j^2)$$
$$l_i^2 \geq l_i^2$$

Clearly, the equality is satisfied. On the other hand, if say $\theta_k > 90°$, $l_k^2 > l_i^2 + l_j^2$ and equation 3.74 becomes

$$l_i^2 \geq \frac{1}{2}(l_k^2 - l_i^2 - l_j^2) + \frac{1}{2}(l_k^2 + l_i^2 - l_j^2)$$
$$l_i^2 \geq l_k^2 - l_j^2$$

which is not satisfied.

Examination of the second and third rows of the typical element stiffness matrix serves to confirm that the equality in the diagonal dominance condition is satisfied for all rows if none of the angles of the element exceeds 90°. While the absence of obtuse-angled elements is sufficient to ensure convergence of the Gauss–Seidel method for problems of the harmonic type, it may not be necessary: the inclusion of some obtuse-angled triangles may be permissible. It is interesting to note that the equivalent finite difference formulation in the form of equation 2.92 also satisfies the equality in the diagonal dominance condition.

3.6.3 Comparison of methods   The gaussian elimination and Gauss–Seidel methods for solving sets of finite element equations can be compared in terms of computing time and storage requirements. The most effective way to compare computing times is to compare the numbers of arithmetic operations. It is necessary to distinguish between divisions, multiplications and additions (or subtractions) because the ratios between the times for these operations depend on the particular computer.

For gaussian elimination applied to a large number, n, of equations, the computing time is dominated by approximately $\frac{1}{3}n^3$ multiplications, and the same number of additions, if the full square stiffness matrix is used. If the rectangular form is used, the computing time is dominated by $\frac{1}{4}nb^2$ multiplications, and the same number of additions. Typically, the matrix bandwidth b is of the order indicated by equation 3.68 and the number of multiplications and additions is of the order of n^2.

In the above version of the Gauss–Seidel method for sparse stiffness matrices, if there are up to p nonzero stiffness coefficients per row then approximately pn multiplications and the same number of additions are required per cycle of iteration. For present purposes, including the computer program described in section 3.8, the value of p is taken to be 9: finite element meshes must be such that no nodal point is directly connected to more than 8 adjacent nodes. If q iterations are required for convergence, the Gauss–Seidel method is faster than the full elimination method if

$$r_1 < 1, \quad r_1 = \frac{3pq}{n^2} = \frac{27q}{n^2} \tag{3.75}$$

and faster than elimination applied to the rectangular form of stiffness matrix if

$$r_2 < 1, \quad r_2 = \frac{pq}{n} = \frac{9q}{n} \tag{3.76}$$

These conditions are examined in the case study presented in section 5.1.

The storage requirement for solving finite element equations is dominated by the stiffness matrix. The full matrix requires n^2 coefficients to be stored, but this number is reduced to bn if the banded form described in section 3.6.1 is used. The Gauss–Seidel method requires only $2pn$ coefficients to be stored in matrices $\bar{K}$ and M discussed in section 3.6.2. Since b and p are generally of the order of $2\sqrt{n}$ and 9 respectively, the Gauss–Seidel method requires substantially less storage than either of the elimination methods.

A method for solving a particular set of finite element equations may be selected according to the size of problem and type of computer to be used, as follows.

(1) For problems involving a small enough number of nodal points for the full stiffness matrix to be contained in the available fast core store of the computer, the basic gaussian elimination method can be employed. This is the most generally applicable method for solving sets of linear equations and its use is demonstrated in the case study described in section 1.3.

(2) For problems involving stiffness matrices too large to be stored in full but small enough to be stored in rectangular form in the fast core store, either gaussian elimination or the Gauss–Seidel method can be used. While the former may be faster, the latter requires less storage, and the choice between them is not always clearcut. Gaussian elimination is normally preferable if the stiffness matrix is not diagonally dominant, but is unsuitable for nonlinear problems of the type discussed in section 8.7. Gaussian elimination is also the better choice if several problems involving the same stiffness matrix but different load vectors are to be solved, since the elimination process need only be applied once to the stiffness matrix.

(3) For problems involving so many nodal points that even the rectangular forms of the stiffness matrix cannot be contained in the fast core store, backing stores in the form of magnetic tapes or discs must be employed. Since these are slow their use should be kept to a minimum. Only part of the rectangularised stiffness matrix can be held in the core at any one time. In the gaussian elimination method the stiffness matrix need only be passed through the core once during the elimination process and once to enable back substitution to be carried out. During elimination the matrix is transferred, one section at a time, from the backing store into the fast store, modified and then returned to the backing store. Since the Gauss–Seidel method would require the stiffness matrix to be passed through the core once every cycle of iteration, it should not be used to solve large problems.

Apart from the simple structural analysis example described in section 1.3, all the practical problems considered in this book fall into the second of the above categories. Attention is concentrated on the Gauss–Seidel method of solution because it is particularly simple to program.

3.7 Convergence of Finite Element Methods

The convergence of the Gauss–Seidel technique for solving sets of linear equations is discussed in section 3.6.2. An even more important and quite distinct form of convergence is that of the finite element method as a whole. In all such methods it is assumed that, as the numbers of elements and nodal points are increased, the computed solution approximates more closely the true solution to the problem, which is unique. For present purposes it can be argued from the principles of continuum mechanics outlined in section 2.1 that this convergence is assured if the computed solution satisfies as closely as it can the conditions of equilibrium and energy conservation, compatibility of strains or strain rates, constitutive relationships and the relevant boundary conditions.

In the case of the downstream flow problem considered in this chapter, equilibrium of nodal point forces, the constitutive equations and the boundary conditions are satisfied explicitly. Energy conservation need only be considered if the temperature distribution in the flow is required. Consequently, the condition required to assure convergence is that of compatibility of strain rates. As indicated in section 3.1.1, this compatibility is automatically satisfied within each element and the velocity is continuous across inter-element boundaries: the method is convergent.

3.7.1 Convergence conditions for the variational formulation   The variational formulation described in section 3.4 provides a more general view of the conditions required for convergence of finite element methods. In equation 3.50 the total value of the functional χ over the whole of the solution domain is equated to the sum of the results of the relevant integrations performed over all the individual elements. This assumes that the interfaces between the elements make no contribution to χ. Figure 3.9 shows two adjacent elements separated along their interface by a small gap h. The condition for this interface to make no contribution to χ is

$$\lim_{h \to 0} \int_0^h \left[\frac{1}{2}\left(\frac{\partial \psi}{\partial x}\right)^2 + \frac{1}{2}\left(\frac{\partial \psi}{\partial y}\right)^2 + \psi\phi_1 \right] ds = 0 \qquad (3.77)$$

If ψ is continuous across the interface, its value remains constant over the gap. The first derivatives of ψ with respect to x and y are therefore zero and equation 3.77 is satisfied. On the other hand, if ψ is not continuous across the interface, its value undergoes an abrupt change over the gap and the magnitudes of the derivatives tend to infinity as h is reduced to zero. Consequently, the limit defined in equation 3.77 is indeterminate, although it may be zero.

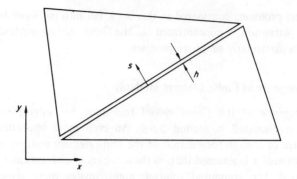

Figure 3.9   Two adjacent elements

The variational formulation provides a general statement of the conditions for convergence of finite element methods. If the functional whose stationary value is sought involves derivatives of the unknown function up to the nth order, the shape functions employed within the elements should ensure continuity across element interfaces of derivatives up to the $(n - 1)$th order. Elements and their associated shape functions which satisfy this requirement are said to be 'conforming' or 'compatible'. Even if they are not conforming, however, the method may still converge, and many such elements have been used successfully.

3.7.2 Truncation errors   That it is necessary to use a large number of elements and nodal points to achieve satisfactory convergence of finite element methods is due to the fact that shape functions such as equation 3.5 provide only approximate representations of the true variations. A Taylor series expansion about the origin of the local co-ordinates shown in figure 3.2 for the velocity gives

$$w = w_i + \left(x\frac{\partial}{\partial x} + y\frac{\partial}{\partial y}\right) w + \frac{1}{2}\left(x\frac{\partial}{\partial x} + y\frac{\partial}{\partial y}\right)^2 w + \ldots \qquad (3.78)$$

the derivatives being evaluated at the origin. Now $C_1 = \partial w/\partial x$, $C_2 = \partial w/\partial y$, and the error involved in using equation 3.5 as a truncated form of equation 3.78 to represent the velocity at the point j, say, is of the order of

$$e_T = \frac{1}{2}\left(a_k^2 \frac{\partial^2 w}{\partial x^2} - 2a_k b_k \frac{\partial^2 w}{\partial x\, \partial y} + b_k^2 \frac{\partial^2 w}{\partial y^2}\right) \qquad (3.79)$$

This truncation error is of the order of the square of the dimensions of the element and tends to zero as the element size is reduced. As is to be expected from the comparison discussed in section 3.3, the order of the error is the same as that of the finite difference method described in section 2.4.1.

3.7.3 Solution bounds   Since finite element methods provide only approximate solutions to problems, it is of interest to consider whether these solutions form upper or lower bounds to the true solutions. The variational formulation provides the most general way of examining solution bounds. The true solution to a problem provides a stationary value for the functional χ. In the case of the functional shown in equation 3.48, the true solution to the downstream flow problem provides the absolute minimum value of χ. It is to be expected, therefore, that the finite element solution corresponds to a value of χ which, although minimised within the contraints of the assumed element shape functions, is greater than the true minimum. For a conforming finite element formulation based on velocities this means that the computed element and overall stiffnesses provide upper bounds to the true stiffnesses. At least for problems in which boundary velocities are prescribed, the computed velocities therefore provide lower bounds for the true values.

3.8   A Computer Program for Harmonic Problems

Having presented in detail a finite element analysis for two-dimensional problems of the harmonic type, it now remains to express the method in the form of a computer program. Some practical applications are described in chapter 5, following chapter 4 which is devoted to the provision of mesh data for such problems. The program is divided into a number of subprograms, both to provide largely independent units for programming purposes and to allow different versions of these units to be used to solve a wide range of problems.

Figure 3.10 shows the main program and figure 3.11 a subprogram named MSHOUT for writing out the geometric data for the mesh of elements and nodal points. Figure 3.12 shows a subprogram named SOLVE1 for applying the Gauss—Seidel method of solving sparse sets of linear equations. Other subprograms called by the main program are MESH and MODIFY which provide the mesh data as described in chapter 4, PHI1F which provides the distribution of the ϕ_1 function occurring in the general harmonic equation 3.40, BCS which provides the boundary conditions for the particular problem, and OUTPUT which writes out the required results. Examples of these last three subprograms are provided by the case studies described in chapter 5.

3.8.1 The main program   The main program shown in figure 3.10 is based on the analysis described in sections 3.1 and 3.2, but is generalised to solve any two-dimensional problem of the harmonic type governed by a differential equation of the form of equation 2.87. Since both sides of equation 2.30 which governs downstream flow are divided by viscosity to give equation 2.31 in the general form of equation 2.87, so in the program viscosity is effectively set to unity in the stiffness analysis and pressure gradient is divided by viscosity to give the function ϕ_1.

```
C   PROGRAM FOR FINITE ELEMENT ANALYSIS OF TWO-DIMENSIONAL PROBLEMS OF
C   THE HARMONIC TYPE, USING CONSTANT STRAIN (RATE) TRIANGULAR ELEMENTS.
C
      DIMENSION  TITLE(8),PHI1(200),B(2,3),ESTIFF(3,3),IJK(3)
      COMMON /CMESH/ NEL,NNP,X(121),Y(121),AI(200),AJ(200),AK(200),
     1    BI(200),BJ(200),BK(200),AREA(200),NPI(200),NPJ(200),NPK(200),
     2    NBP,NPB(40),MOUT
     3          /CEQNS/ OSTIFF(121,9),DELTA(121),F(121),NPA(121,9),NAP(121)
      DATA  BLANK / 10H          /
C
C   INPUT THE PROBLEM TITLE - BUT STOP IF BLANK CARD ENCOUNTERED.
  1   READ(5,51) TITLE
 51   FORMAT(8A10)
      IF(TITLE(1).EQ.BLANK) STOP
      WRITE(6,61) TITLE
 61   FORMAT(65HOCST FINITE ELEMENT SOLUTION FOR TWO-DIMENSIONAL HARMONI
     1C PROBLEM // 8A10)
C
C   INPUT OR GENERATE THE MESH DATA AND PHI FUNCTION (FOR THE HARMONIC
C   EQUATION).
      CALL   MESH
      CALL   MODIFY
      CALL   PHI1F(PHI1,NEL)
C
C   COMPUTE THE ELEMENT GEOMETRIES.
      DO 2 M=1,NEL
      I=NPI(M)
      J=NPJ(M)
      K=NPK(M)
      AI(M)=-X(J)+X(K)
      AJ(M)=-X(K)+X(I)
      AK(M)=-X(I)+X(J)
      BI(M)=Y(J)-Y(K)
      BJ(M)=Y(K)-Y(I)
      BK(M)=Y(I)-Y(J)
      AREA(M)=0.5*(AK(M)*BJ(M)-AJ(M)*BK(M))
      IF(AREA(M).GT.0.) GO TO 2
      WRITE(6,62) M
 62   FORMAT(15HOELEMENT NUMBER,I5,25H HAS NEGATIVE AREA - STOP)
      STOP
  2   CONTINUE
C
C   OUTPUT THE MESH GEOMETRY DATA.
      CALL MSHOUT
C
C   SET INITIAL VALUES OF STIFFNESSES, EXTERNAL FORCES AND UNKNOWNS.
      DO 4 IROW=1,NNP
      DO 3 IC=1,9
      OSTIFF(IROW,IC)=0.
  3   NPA(IROW,IC)=0
      NPA(IROW,1)=IROW
      F(IROW)=0.
  4   DELTA(IROW)=0.
C
C   SET UP THE OVERALL ASSEMBLY LOOP.
      DO 9 M=1,NEL
C
C   STORE THE ELEMENT NODE NUMBERS IN ORDER IN ARRAY IJK.
      IJK(1)=NPI(M)
      IJK(2)=NPJ(M)
      IJK(3)=NPK(M)
C
C   COMPUTE THE EXTERNAL FORCE COMPONENTS ON EACH NODE OF THE ELEMENT.
      FM=-PHI1(M)*AREA(M)/3.
```

Figure 3.10   Main program for finite element analysis of harmonic problems

```
C
C   FORM THE ELEMENT STIFFNESS MATRIX.
       B(1,1)=BI(M)
       B(1,2)=BJ(M)
       B(1,3)=BK(M)
       B(2,1)=AI(M)
       B(2,2)=AJ(M)
       B(2,3)=AK(M)
       FACT=0.25/AREA(M)
       DO 8 IRE=1,3
       DO 8 ICE=1,3
       ESTIFF(IRE,ICE)=FACT*(B(1,IRE)*B(1,ICE)+B(2,IRE)*B(2,ICE))
C
C   ADD ELEMENT STIFFNESS TO OVERALL STIFFNESS.
       IROW=IJK(IRE)
       ICOL=IJK(ICE)
C
C   STORE STIFFNESS COEFFICIENT IN RECTANGULAR FORM OF OVERALL MATRIX.
       DO 5 IC=1,9
       IF(NPA(IROW,IC).EQ.ICOL) GO TO 7
       IF(NPA(IROW,IC).EQ.0) GO TO 6
   5   CONTINUE
       WRITE(6,63) IROW
   63  FORMAT(5H0NODE,I5,38H HAS MORE THAN 8 ADJACENT NODES - STOP)
       STOP
   6   NPA(IROW,IC)=ICOL
       NAP(IROW)=IC
   7   OSTIFF(IROW,IC)=OSTIFF(IROW,IC)+ESTIFF(IRE,ICE)
   8   CONTINUE
C
C   ASSEMBLE THE EXTERNAL FORCES ON THE NODES.
       DO 9 IRE=1,3
       IROW=IJK(IRE)
   9   F(IROW)=F(IROW)+FM
C
C   APPLY THE BOUNDARY CONDITIONS.
       CALL   BCS
C
C   SOLVE THE LINEAR EQUATIONS.
       CALL   SOLVE1(NNP)
C
C   OUTPUT THE REQUIRED RESULTS.
       CALL   OUTPUT
       GO TO 1
       END
```

Figure 3.10   Continued

As far as possible the program variable names are the same as those used in the other programs presented in this book. The geometric data for the mesh are stored in arrays which are located in the COMMON block of storage named CMESH. The numbers of elements and nodal points are stored in NEL and NNP respectively, while the global co-ordinates of the nodal points are stored in order of node number in the arrays X and Y. The arrays AI, AJ, AK, BI, BJ and BK are used to store, in order of element number, the element dimensions a_i, a_j, a_k, b_i, b_j and b_k shown in figure 3.2. Element areas are stored in the array AREA, while NPI, NPJ and NPK store the numbers of the nodal points (i, j and k for the typical element shown in figure 3.2) at the corners of the elements. The array NPB stores the numbers assigned to the nodal points located on the boundary of the solution domain, and NBP stores the total number of such points. The variable MOUT stores an integer parameter which is used to control the amount of mesh data written out by subprogram MSHOUT. In the present form of the program, the dimensions of the arrays storing the mesh data are such as to allow up to 200 elements, 121 nodal points and 40 boundary points to be used. This corresponds to, for example, a rectangular mesh involving 11 rows of 11 nodes, each of the 100 small rectangles so formed being divided into two triangular elements. Meshes of this type are discussed in section 4.3.1.

The arrays of variables used in the solution of the overall set of linear equations are located in the COMMON block of storage named CEQNS. Arrays OSTIFF, NPA and NAP are used to store the overall stiffness data in the rectangular form illustrated by matrices $\widetilde{K}$, M and L shown in equations 3.70. The actual stiffness coefficients are stored in OSTIFF, while NPA stores the numbers assigned to the nodal points adjacent to each of the nodes, and NAP stores the total numbers of such points. In this context, adjacent points include the point itself and the self-stiffness coefficients are stored in the first column of OSTIFF. The dimensions of OSTIFF and NPA are such as to allow the use of meshes in which no nodal point is directly connected to more than eight other adjacent nodes. The arrays DELTA and F are used to store, in order of node numbers, the nodal point variables such as velocities or displacements, and the corresponding externally applied forces: that is, the coefficients of vectors δ and F shown in equation 3.23.

The other subscripted variables used are TITLE which stores an alphanumeric title for the problem, PHI1 which stores for each element the mean value of the function ϕ_1, B and ESTIFF which store the coefficients of the element dimension and stiffness matrices B and k_m, and IJK which stores in order the node numbers i, j, k for a particular element. Other variables used in the main program include I, J and K for nodal point numbers and M for element numbers. IROW and ICOL are used to store row and column numbers for the full (square) overall stiffness matrix, while IRE and ICE serve the same purpose for the element stiffness matrices (p, q, r and s are the equivalent counters in equations 3.26 and 3.27). IC is used to contain the column number in the rectangular form of the overall stiffness matrix.

The first action of the main program is to read in the problem title from the first data card. If at least the first ten columns of this card are blank then execution is terminated. Otherwise a heading is written out, followed by the problem title. Then subprograms MESH, MODIFY and PHI1F are called in turn to provide the mesh data and ϕ_1 function distribution.

Since subprograms MESH and MODIFY serve only to define the co-ordinates of the nodal points and the nodes associated with each element, the geometries of the elements are computed in the main program. For each element the numbers assigned to the three nodes are obtained from the arrays NPI, NPJ and NPK, and the element dimensions and area found with the aid of equations 3.1 and 3.4. If a negative element area is encountered, which is normally due to the three nodes being numbered in clockwise rather than anticlockwise order, execution is terminated. The subprogram MSHOUT is then called to write out the geometric data for the mesh.

The coefficients stored in arrays OSTIFF, NPA and F are set to zero in preparation for the assembly of the overall stiffness matrix and external force vector. The first column of OSTIFF is made to contain the self-stiffness coefficients of the nodal points by setting the node counters stored in the first column of NPA equal to the corresponding row numbers. The unknowns stored in array DELTA are also set to zero to serve as the initial values for the Gauss–Seidel solution process.

Within the program loop for the overall assembly process the external force components and stiffness matrix are computed for each element in turn. The node numbers for the particular element are first stored in anticlockwise order in array IJK. This provides the relationships between the row and column numbers in the element and overall stiffness matrices required in equation 3.26. For example, the row number p in this equation can be obtained as the rth component of the array IJK. The external force components acting on each node of the element are found with the aid of equation 3.14 and stored in the variable FM. Note that the pressure gradient appearing in equation 3.14 is replaced in the program by P_z/μ, or in general the mean value of the function ϕ_1 for the particular element. The coefficients of the dimension matrix B are computed according to equation 3.12 and the coefficients of the element stiffness matrix found with the aid of equation 3.20 as

$$k_{rs} = \frac{\mu}{4\Delta_m} (B_{1r} B_{1s} + B_{2r} B_{2r}) \tag{3.80}$$

where, for example, B_{1r} is the coefficient in the first row and rth column of matrix B. As already explained, in the program the viscosity is set to unity in the stiffness analysis, and the variable FACT serves to store the common factor $\frac{1}{4}\Delta_m$.

In order to add a particular element stiffness coefficient to the overall stiffness matrix, the corresponding row and column numbers in the full (square) stiffness matrix are first obtained from array IJK and stored in IROW and ICOL respectively. Since the overall stiffness matrix is stored in rectangular form a search is made of the appropriate row of NPA to determine whether the current value of ICOL is already stored: that is, whether the same pair of nodal points has already occurred in a previously analysed element. If it is already stored the element stiffness coefficient is added to the corresponding coefficient of OSTIFF. If it is not already stored the search of NPA will eventually yield a zero in the first unused column, in which the value of ICOL is then stored. The element stiffness coefficient is added to the corresponding (zero) coefficient of OSTIFF and the current number of nonzero coefficients in the particular row is stored in NAP. If the mesh is such that the maximum permitted number of nodes adjacent to a particular point is exceeded, no vacant column of NPA can be found and execution is terminated. This process of storing overall stiffness coefficients in rectangular array form can perhaps be more readily understood with the aid of equations 3.69 and 3.70. The order of storage of coefficients along rows of OSTIFF after the first column depends on the order in which the elements are numbered.

The last action within the overall assembly loop is to add the external nodal point force components due to the presence of each element to the total external forces acting on the relevant nodes. When the overall assembly is complete, the boundary conditions are applied by subprogram BCS. The set of linear equations is solved in subprogram SOLVE1 and the results written out in subprogram OUTPUT. Then control of execution in the main program returns to the first input statement for a new problem.

```
      SUBROUTINE  MSHOUT
C
C  SUBPROGRAM TO WRITE OUT THE GEOMETRIC DATA FOR THE MESH.
C
      COMMON /CMESH/ NEL,NNP,X(121),Y(121),AI(200),AJ(200),AK(200),
     1      BI(200),BJ(200),BK(200),AREA(200),NPI(200),NPJ(200),NPK(200),
     2      NBP,NPB(40),MOUT
      IF(MOUT.EQ.0) RETURN
C
C  OUTPUT THE NUMBER OF ELEMENTS, NODAL POINTS AND CO-ORDINATES.
      WRITE(6,61) NEL,NNP,(I,X(I),Y(I),I=1,NNP)
   61 FORMAT(28HOGEOMETRIC DATA FOR THE MESH //
     1  10X,21H NUMBER OF ELEMENTS =,I4 //
     2  10X,25H NUMBER OF NODAL POINTS =,I4 //
     3       25H NODAL POINT CO-ORDINATES //
     4  72H  I      X        Y       I       X       Y      I
     5  X        Y     / (3(1X,I5,2F9.4)))
C
C  OUTPUT THE ELEMENT NODE NUMBERS AND AREAS.
      WRITE(6,62) (M,NPI(M),NPJ(M),NPK(M),AREA(M),M=1,NEL)
   62 FORMAT(31HOELEMENT NODE NUMBERS AND AREAS //
     1  66H   M    I    J    K       AREA        M    I    J    K
     2  AREA      / (2(1X,4I5,E12.4)))
      RETURN
      END
```

Figure 3.11   Subprogram for writing out mesh data

3.8.2 The mesh data output subprogram   The subprogram MSHOUT shown in figure 3.11 serves to write out the geometric data for the mesh. Following the number of elements and nodal points, the numbers and global co-ordinates of the nodal points are written out, to be printed three sets to a line. Then the element numbers, node numbers and areas are written out, to be printed two sets to a line. Note that, while for the purposes of this book the FORMAT statements used restrict output to no more than 72 characters per line, in general the full width of paper on the line printer should be used. With the subprogram in its present form, the variable MOUT provides the only means of controlling the amount of mesh data printed out. None or all of the data mentioned above are printed according to whether the value of this parameter is zero or nonzero. The required value is defined in subprogram MESH.

3.8.3 The Gauss–Seidel subprogram for harmonic problems   The subprogram shown in figure 3.12 is given the name SOLVE1 to distinguish it from the similar one named SOLVE2 which is introduced later in connection with biharmonic problems. All the equation data required by the subprogram are obtained from the COMMON block of storage named CEQNS, with the exception of the

```
      SUBROUTINE  SOLVE1(NNP)
C
C  SUBPROGRAM FOR SOLVING BY GAUSS-SEIDEL METHOD THE LINEAR EQUATIONS
C  OBTAINED FROM THE FINITE ELEMENT FORMULATION OF HARMONIC PROBLEMS.
C
      COMMON /CEQNS/ OSTIFF(121,9),DELTA(121),F(121),NPA(121,9),NAP(121)
      NEQN=NNP
C
C  INPUT THE SOLUTION PARAMETERS.
      READ(5,51) NCYCLE,IFREQ,ORELAX,TOLER
   51 FORMAT(2I5,2F10.0)
      WRITE(6,61) ORELAX
   61 FORMAT(48HOSOLUTION OF EQUATIONS BY GAUSS-SEIDEL ITERATION //
     1         25H OVER-RELAXATION FACTOR =,F6.3)
C
C  SET UP ITERATION LOOP.
      IF(IFREQ.NE.0) WRITE(6,62)
   62 FORMAT(21H  ITER       ERROR   )
      DO 3 ITER=1,NCYCLE
      SUMD=0.
      SUMDD=0.
C
C  OBTAIN NEW ESTIMATE FOR EACH UNKNOWN IN TURN.
      DO 2 IROW=1,NEQN
      DELD=F(IROW)
      ICMAX=NAP(IROW)
      DO 1 IC=1,ICMAX
      ICOL=NPA(IROW,IC)
   1  DELD=DELD-OSTIFF(IROW,IC)*DELTA(ICOL)
      DELD=DELD/OSTIFF(IROW,1)
      SUMDD=SUMDD+ABS(DELD)
      DELTA(IROW)=DELTA(IROW)+DELD*ORELAX
   2  SUMD=SUMD+ABS(DELTA(IROW))
C
C  TEST FOR CONVERGENCE.
      ERROR=SUMDD/SUMD
      IF(ERROR.LT.TOLER) GO TO 4
C
C  OUTPUT PROGRESS INFORMATION EVERY IFREQ CYCLES, UNLESS IFREQ=0.
      IF(IFREQ.EQ.0) GO TO 3
      IF(MOD(ITER,IFREQ).EQ.0) WRITE(6,63) ITER,ERROR
   63 FORMAT(1X,I5,E15.4)
   3  CONTINUE
C
C  NORMAL EXIT FROM ITERATION LOOP INDICATES FAILURE TO CONVERGE.
      WRITE(6,64) NCYCLE
   64 FORMAT(21HONO CONVERGENCE AFTER,I5,7H CYCLES)
      RETURN
C
C  OUTPUT NUMBER OF ITERATIONS AND TOLERANCE FOR CONVERGED SOLUTION.
   4  WRITE(6,65) TOLER,ITER
   65 FORMAT(38HOITERATION CONVERGED TO A TOLERANCE OF,E12.4,
     1         6H AFTER,I5,7H CYCLES)
      RETURN
      END
```

Figure 3.12   Subprogram for applying the Gauss–Seidel method

number of nodal points which is entered as an argument. The value of this argument is immediately assigned to the local variable NEQN, which therefore seves to define the number of equations to be solved.

The iteration control parameters are read in: the maximum number of cycles into NCYCLE, output frequency into IFREQ, over-relaxation factor into ORELAX and convergence tolerance into TOLER. If the value of IFREQ is ten, for example, data indicating the progress of the iteration will be written out after every ten cycles. The value of the over-relaxation factor is written out, followed by a heading for the iteration progress data. A zero value of IFREQ implies that such data and therefore the heading are not required.

The iteration loop is set up with ITER as a counter. The variables SUMD and SUMDD, which are used to accumulate the summed magnitudes of δ_i and $\Delta\delta_i$ required for the convergence test of the form defined in equation B.2, are set to zero at the beginning of each iteration. For each equation in turn, the variable DELD is used to accumulate the value of $\Delta\delta_i$ as defined by equation 3.72, from which the new value of δ_i is obtained with the aid of the over-relaxation factor and equation B.5.

At the end of each iteration cycle, the relative error is calculated according to equation B.2 and stored in the variable ERROR. If the convergence criterion is satisfied for the tolerance stored in TOLER, then the value of this tolerance and the number of iterations are written out, and the converged solutions are returned to the calling program. Otherwise, the iteration number and relative error are written out whenever ITER is divisible by the nonzero value of IFREQ. If the iteration loop is completed by the value of ITER reaching that of NCYCLE, then convergence to the required tolerance is not achieved, and a warning message is written out before control of execution is returned to the calling program.

4   Finite Element Meshes

In chapter 3 a finite element analysis for harmonic problems is developed to the point of presenting a computer program, but without discussing in detail how the data required to define the mesh of elements are to be provided. Since the program for biharmonic problems described in chapter 6 requires mesh data in identical forms, it is convenient to consider the provision of such data independently of problem type. In this chapter attention is confined to meshes of two-dimensional triangular elements with nodal points at their corners. Chapters 5 and 7 provide a number of practical examples of the use of such meshes in problems of the harmonic and biharmonic types respectively.

In the finite element programs described in sections 3.8 and 6.6, mesh data are provided with the aid of two subprograms, named MESH and MODIFY. MESH serves to establish the basic mesh geometry and numbering of the nodal points and elements by either reading in the relevant data as described in section 4.2, or using the computer to generate the data as described in section 4.3. Subsequently, MODIFY is used to adapt the basic mesh to the requirements of the particular problem, as described in section 4.4. Before considering examples of these subprograms, however, it is appropriate to review the criteria for designing finite element meshes.

4.1   Choice of Mesh

Meshes of triangular elements can be devised to suit a very wide range of practical problems. The first requirement of such a mesh is that it should fit the shape of the boundary of the solution domain as closely as possible. This requirement can be met provided the boundary shape can be approximated with sufficient accuracy by a series of short straight lines which form sides of elements. Clearly, the sizes of the elements should be reduced in regions close to sharply curved boundaries. Following the same principle within the solution domain, since the variations of the unknowns are assumed to be linear over each element it is desirable to have a concentration of relatively small elements in regions of the domain where the unknowns are likely to change rapidly, particularly since these regions are often of greatest interest in the solution. Especially in the case of biharmonic problems, however, element sizes should not change too rapidly in such regions. The case study described in section 7.3 provides a practical example of this type of situation. As for the shape of individual elements, for harmonic problems the main consideration is the avoidance of obtuse angles, particularly if the Gauss–Seidel method of solution is to be used (see section 3.6.2). For biharmonic problems the requirements are rather more restrictive and ideally the elements should be as nearly equilateral as possible (see section 6.4).

In addition to element size, shape and distribution within the mesh, the numbering of both elements and nodal points needs to be considered. The order of numbering elements can be arbitrary, although in practice it is convenient to arrange for it to follow a regular pattern. The requirements for ordering of node numbers can be slightly more restrictive. If gaussian elimination is to be used to solve the overall linear equations, then as indicated in section 3.6.1 it is important to try to minimise the bandwidth of the overall stiffness matrix. To find the absolute minimum value of this bandwidth for a complicated mesh can be a major computational task in itself. Nevertheless, this requirement should be borne in mind when numbering nodes. For example, in the square mesh shown in figure 4.2 the nodes are numbered from left to right row by row. For the particular mesh the resulting bandwidth is minimal, although if there were fewer nodes in each vertical row than in each horizontal one it would be better to number them in vertical rows, say from top to bottom within each row.

Finally, having located and numbered the nodal points it is necessary to define which three nodes form the corners of each element. The numbers of these nodes should be defined in an anticlockwise order, although it is immaterial which node is defined first and therefore forms the local co-ordinate origin for the element. If a clockwise order of numbering is inadvertently used, the element area calculated according to equation 3.4 is negative. This provides a useful means of checking the general validity of a particular set of mesh data, which is used in the programs for both harmonic and biharmonic problems. Serious errors in the data nearly always result in one or more elements having zero or negative areas.

4.2   Mesh Data in Numerical Form

One way of providing the mesh data required by a finite element program is to enter them on data cards. The structural analysis case study described in section 1.3 provides an illustration of this type of approach, and figure 4.1 shows a suitable form of the subprogram MESH for reading in the data required by the programs for harmonic and biharmonic problems. The total numbers of elements and nodal points are first read into variables NEL and NNP respectively,

```
      SUBROUTINE  MESH
C
C   SUBPROGRAM TO READ OR GENERATE A MESH OF TRIANGULAR FINITE ELEMENTS.
C   THIS VERSION READS IN THE NECESSARY DATA.
C
      COMMON /CMESH/ NEL,NNP,X(121),Y(121),AI(200),AJ(200),AK(200),
    1      BI(200),BJ(200),BK(200),AREA(200),NPI(200),NPJ(200),NPK(200),
    2      NBP,NPB(40),MOUT
C
C   INPUT THE NUMBERS OF NODES AND ELEMENTS, ALSO THE MESH DATA OUTPUT
C   CONTROL PARAMETER.
      READ(5,51) NNP,NEL,MOUT
   51 FORMAT(4I5)
      IF(NNP.LE.121.AND.NEL.LE.200) GO TO 1
      WRITE(6,61) NNP,NEL
   61 FORMAT(30HOEXCESSIVE SIZE OF MESH, NNP =,I5,8H,   NEL =,I5)
      STOP
C
C   INPUT THE NODAL POINT CO-ORDINATES.
    1 READ(5,52) (I,X(I),Y(I),N=1,NNP)
   52 FORMAT(I5,2E15.5)
C
C   INPUT THE ELEMENT NODE DATA.
      READ(5,51) (M,NPI(M),NPJ(M),NPK(M),N=1,NEL)
      RETURN
      END
```

Figure 4.1   Subprogram for reading in mesh data

together with the value of the mesh data output control parameter into MOUT, the use of which is described in section 3.8.2. Then the global co-ordinates of the nodal points are read into the arrays X and Y, the node number being read into the temporary store I in each case: the cards supplying these data can be arranged in arbitrary order. It is also possible to supply initial estimates of the values of the unknowns at the nodal points to serve as starting values for the Gauss–Seidel solution process. Good estimates are rarely known, however, and as indicated in appendix B their use often has comparatively little effect on the number of iterations required for convergence. Following the node data, the element data in the form of the numbers assigned to the three nodes of each triangle are read into arrays NPI, NPJ and NPK, the element number being read into the temporary store M in each case. Other data such as material properties for the elements could also be read in from the same cards.

It should be noted that the number of data cards for even relatively coarse meshes tends to become very large. For example, consider a square mesh of the form shown in figure 4.2, but with 11 rows of 11 nodes. This would involve 121 nodal points and 200 elements, the maximum permitted by the present array dimensions, and a total of over 320 data cards if subprogram MESH shown in figure 4.1 is used. A considerable amount of labour is involved in preparing these cards and since mistakes are likely to be made the data must be carefully checked. One useful test is that for zero or negative element areas described in section 4.1. Another is to use a mesh plotting program and the graph plotting facility available with most computers to draw out the mesh and number the nodes and elements. Errors in the nodal point co-ordinates and the element node numbers are readily apparent from the drawn mesh. Various mechanical aids have been developed to help to automate the process of translating mesh geometry into numbers punched on data cards.

The main advantage of providing mesh data in numerical form is that maximum flexibility of mesh geometry is possible. The mesh can be specified in precisely the form thought to be the best choice for a particular problem. The disadvantages in terms of the quantity of input data and the likelihood of errors are often sufficient, however, to encourage the search for a method of generating the mesh data within the program.

4.3   Generation of Mesh Data

Ideally, a finite element computer program should generate its own mesh data from a minimum number of geometric parameters. The amount of input data required is minimal and once the relevant coding has been written and tested the possibility of errors is largely eliminated. On the other hand, it may be difficult to devise a suitable algorithm for mesh generation, and few attempts have been made to develop general methods. Particular forms of meshes can, however, be readily generated for simple boundary shapes and can then be modified to suit a wide range of problems. This section is concerned with the generation of some simple meshes, while section 4.4 deals with their subsequent modification.

4.3.1 A square mesh of right-angled triangles   Figure 4.2 shows a square mesh containing six rows of six uniformly spaced nodal points and a total of fifty right-angled triangular elements. Considering the general case where there are n_x points per horizontal row and n_y points per vertical row, the total numbers of nodes and elements are $n_x n_y$ and $2(n_x - 1)(n_y - 1)$ respectively. Let i_x be used to count nodes from left to right in a particular horizontal row and i_y be used to count such rows from the bottom to the top of the mesh, where $1 \leqslant i_x \leqslant n_x$ and $1 \leqslant i_y \leqslant n_y$. If the order of node numbering is from left to right along horizontal rows taken in order from bottom to top as shown, then the number of a typical node can be obtained from the corresponding values of the counters as

$$i = (i_y - 1)n_x + i_x \tag{4.1}$$

Assuming the mesh has unit overall dimensions in the horizontal and vertical directions and that the origin of the global co-ordinates is at the first node, the co-ordinates of the typical point are

$$X_i = \frac{i_x - 1}{n_x - 1}, \quad Y_i = \frac{i_y - 1}{n_y - 1} \tag{4.2}$$

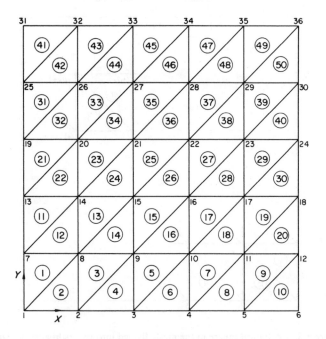

Figure 4.2 A square mesh of right-angled triangles

The numbers of the nodal points at the corners of each element may be defined by considering the overall mesh to be divided into a total of $(n_x - 1)(n_y - 1)$ small squares, each one of which is then subdivided into two triangles. Figure 4.3 shows a typical square and pair of elements. Let i_x and i_y be horizontal and vertical counters as before but this time applied to squares, where now $1 \leqslant i_x \leqslant n_x - 1$ and $1 \leqslant i_y \leqslant n_y - 1$. If the squares are considered to be numbered from left to right along horizontal rows taken in order from bottom to top, then the number of a particular square can be obtained from the corresponding values of the counters as

$$n_q = (i_y - 1)(n_x - 1) + i_x \tag{4.3}$$

Since there are twice as many elements as there are squares, the numbers of the two elements into which this square is subdivided are

$$m_1 = 2n_q - 1, \quad m_2 = 2n_q \tag{4.4}$$

The number of the node i at the bottom left-hand corner of the square is given by equation 4.1. If this node is taken as the local origin for both elements, then as shown in figure 4.3 their nodes taken in anticlockwise order are

$$i_1 = i, \quad j_1 = i + n_x + 1, \quad k_1 = i + n_x$$
$$i_2 = i, \quad j_2 = i + 1, \quad k_2 = i + n_x + 1 \tag{4.5}$$

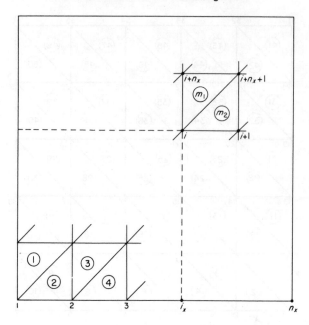

Figure 4.3　A typical square in the mesh divided into two right-angled triangles

Figure 4.4 shows a version of subprogram MESH for generating mesh data by the above method. The variables NXPT, NYPT, IX and IY are used to store the values of n_x, n_y, i_x and i_y respectively, while I, NSQ, M1 and M2 serve to store the values of i, n_q, m_1 and m_2. The variables NXEL and NYEL store the numbers of pairs of elements forming squares in each horizontal and vertical row respectively, that is $n_x - 1$ and $n_y - 1$. Note that the mesh parameters NXPT and NYPT, whose values constitute the only geometric input data, are located in the COMMON block of storage named CMPAR. This makes them accessible to other subprograms.

4.3.2 A square mesh of mainly isosceles triangular elements　While meshes derived from a square one of the form shown in figure 4.2 can be used in a range of applications, it is sometimes desirable to employ elements which are more nearly equilateral in shape. This is particularly important in the case of biharmonic problems, and in general may provide a means of avoiding obtuse-angled elements in a subsequent mesh modification. The modification described in section 7.3 provides an example of such a situation. Figure 4.5 shows a square mesh (again with unit overall dimensions) in which most of the elements are in the form of isosceles triangles. The number of horizontal rows of nodal points, n_y, is shown as five. The number of points along these rows alternate between n_x on the odd-numbered rows such as the bottom one, and $n_x + 1$ on the even-numbered rows. In figure 4.5 the value of n_x is five.

```
      SUBROUTINE   MESH
C
C  SUBPROGRAM TO READ OR GENERATE A MESH OF TRIANGULAR FINITE ELEMENTS.
C  THIS VERSION GENERATES A SQUARE MESH OF RIGHT-ANGLED TRIANGLES.
C
      COMMON /CMESH/ NEL,NNP,X(121),Y(121),AI(200),AJ(200),AK(200),
     1     BI(200),BJ(200),BK(200),AREA(200),NPI(200),NPJ(200),NPK(200),
     2     NBP,NPB(40),MOUT
     3     /CMPAR/ NXPT,NYPT
C
C  INPUT THE NUMBER OF POINTS REQUIRED IN THE X AND Y DIRECTIONS, ALSO
C  THE MESH DATA OUTPUT CONTROL PARAMETER.
      READ(5,51) NXPT,NYPT,MOUT
   51 FORMAT(3I5)
C
C  COMPUTE AND TEST THE NUMBERS OF NODES AND ELEMENTS.
      NNP=NXPT*NYPT
      NEL=(NXPT-1)*(NYPT-1)*2
      IF(NNP.LE.121.AND.NEL.LE.200) GO TO 1
      WRITE(6,61) NNP,NEL
   61 FORMAT(30HOEXCESSIVE SIZE OF MESH, NNP =,I5,8H   NEL =,I5)
      STOP
C
C  DEFINE THE NODAL POINT CO-ORDINATES.
    1 DO 2 IY=1,NYPT
      DO 2 IX=1,NXPT
      I=(IY-1)*NXPT+IX
      X(I)=FLOAT(IX-1)/FLOAT(NXPT-1)
    2 Y(I)=FLOAT(IY-1)/FLOAT(NYPT-1)
C
C  DEFINE THE NUMBERS OF THE THREE NODES OF EACH ELEMENT.
      NXEL=NXPT-1
      NYEL=NYPT-1
      DO 3 IY=1,NYEL
      DO 3 IX=1,NXEL
      NSQ=(IY-1)*NXEL+IX
      M1=NSQ*2-1
      M2=M1+1
      I=(IY-1)*NXPT+IX
      NPI(M1)=I
      NPJ(M1)=I+NXPT+1
      NPK(M1)=I+NXPT
      NPI(M2)=I
      NPJ(M2)=I+1
    3 NPK(M2)=I+1+NXPT
      RETURN
      END
```

Figure 4.4   Subprogram for generating a square mesh of right-angled triangles

Each pair of horizontal rows of points contains $2n_x + 1$ nodes. Hence, if the value of n_y is even the total number of nodes is $n_y(2n_x + 1)/2$ while if it is odd the total is $(n_y - 1)(2n_x + 1)/2 + n_x$. There are $2n_x - 1$ elements per horizontal row, giving a total of $(n_y - 1)(2n_x - 1)$ elements. Let i_x be used to count nodes from left to right in a particular horizontal row and i_y be used to count such rows from the bottom to the top of the mesh. If the value of i_y is an odd number the global co-ordinates of the current node are given by equations 4.2. If it is an even number, however, for $i_x > 1$ the value of X_i must be reduced by an amount equal to half the width of a full element, and for the last point in the row $X_i = 1$.

Since the order of numbering elements can be arbitrary, it is convenient to consider first all those elements whose horizontal sides are lowest. Elements

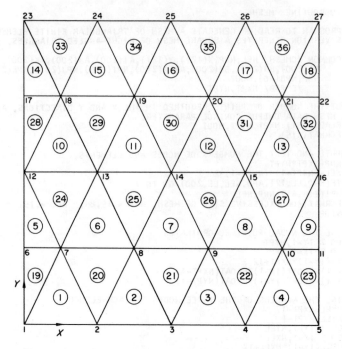

Figure 4.5   A square mesh of mainly isosceles triangular elements

numbered 1 to 18 in figure 4.5 are of this upright type. Let i_y now be used to count horizontal rows of elements from the bottom to the top of the mesh, where $1 \leqslant i_y \leqslant n_y - 1$. Also, let the bottom left-hand corner of each element be the local origin. For the typical upright element numbered m, the numbers of its nodes may be defined in anticlockwise order as

$$i = m + i_y - 1, \quad j = i + 1, \quad k = j + n_x \tag{4.6}$$

Let the total number of upright elements be m_1.

Inverted elements such as those numbered 19 to 36 in figure 4.5 are then considered. Let i_y again be used to count horizontal rows of elements and let the bottom corner of each element be the local origin. For the typical inverted element numbered m, the numbers of its nodes may be defined as

$$i = m - m_1 + i_y - 1, \quad j = i + n_x + 1, \quad k = j - 1 \tag{4.7}$$

Figure 4.6 shows a version of subprogram MESH for generating mesh data by the above method. The variables NXPT, NYPT, IX and IY are used to store the values of n_x, n_y, i_x and i_y respectively, while I, M and M1 serve to store the values of i, m and m_1. NYEL stores the number of horizontal rows of elements, while NXEL stores the number of the current type of element in the current horizontal row. The value of the latter depends on whether the row number is odd or even and

```
      SUBROUTINE   MESH
C
C  SUBPROGRAM TO READ OR GENERATE A MESH OF TRIANGULAR FINITE ELEMENTS.
C  THIS VERSION GENERATES A SQUARE MESH OF MAINLY ISOSCELES ELEMENTS.
C
      COMMON /CMESH/ NEL,NNP,X(121),Y(121),AI(200),AJ(200),AK(200),
     1     BI(200),BJ(200),BK(200),AREA(200),NPI(200),NPJ(200),NPK(200),
     2     NBP,NPB(40),MOUT
     3        /CMPAR/ NXPT,NYPT
C
C  INPUT THE NUMBERS OF POINTS ALONG THE X AND Y AXES, ALSO THE MESH
C  DATA OUTPUT CONTROL PARAMETER.
      READ(5,51) NXPT,NYPT,MOUT
   51 FORMAT(3I5)
C
C  COMPUTE AND TEST THE NUMBERS OF NODES AND ELEMENTS.
      MODNY=MOD(NYPT,2)
      IF(MODNY.EQ.0) NNP=NYPT*(2*NXPT+1)/2
      IF(MODNY.EQ.1) NNP=(NYPT-1)*(2*NXPT+1)/2+NXPT
      NEL=(NYPT-1)*(2*NXPT-1)
      IF(NNP.LE.121.AND.NEL.LE.200) GO TO 1
      WRITE(6,61) NNP,NEL
   61 FORMAT(30HOEXCESSIVE SIZE OF MESH, NNP =,I5,8H,  NEL =,I5)
      STOP
C
C  DEFINE THE NODAL POINT CO-ORDINATES.
    1 I=0
      DO 3 IY=1,NYPT
      MODIY=MOD(IY,2)
      DO 2 IX=1,NXPT
      I=I+1
      X(I)=FLOAT(IX-1)/FLOAT(NXPT-1)
      Y(I)=FLOAT(IY-1)/FLOAT(NYPT-1)
    2 IF(MODIY.EQ.0.AND.IX.GT.1) X(I)=X(I)-0.5/FLOAT(NXPT-1)
      IF(MODIY.EQ.1) GO TO 3
      I=I+1
      Y(I)=Y(I-1)
      X(I)=1.
    3 CONTINUE
C
C  DEFINE THE NUMBERS OF THE THREE NODES OF EACH ELEMENT.
      M=0
      NYEL=NYPT-1
      DO 4 IY=1,NYEL
      NXEL=NXPT-1
      IF(MOD(IY,2).EQ.0) NXEL=NXPT
      DO 4 IX=1,NXEL
      M=M+1
      NPI(M)=M+IY-1
      NPJ(M)=NPI(M)+1
    4 NPK(M)=NPJ(M)+NXPT
      M1=M
      DO 5 IY=1,NYEL
      NXEL=NXPT
      IF(MOD(IY,2).EQ.0) NXEL=NXPT-1
      DO 5 IX=1,NXEL
      M=M+1
      NPI(M)=M-M1+IY-1
      NPJ(M)=NPI(M)+NXPT+1
    5 NPK(M)=NPJ(M)-1
      RETURN
      END
```

Figure 4.6   Subprogram for generating a square mesh of mainly isosceles triangular elements

on whether upright or inverted elements are being considered. MODNY stores the value zero if n_y is an even number, and one if it is odd: MODIY does the same for i_y. The general layout of the subprogram is very similar to that of figure 4.4. Having computed and tested the total numbers of nodes and elements in the mesh, the co-ordinates of the nodal points and the node numbers of the elements are defined as described above. Note that the current values of both the node and element numbers are obtained by simply adding one for each new point or element considered.

4.3.3 A triangular mesh of equilateral elements   Another basic geometric shape of mesh which is often useful is a triangle. Triangular elements are particularly well suited to forming such a mesh, which for the purposes of data generation can be assumed to be equilateral with sides of unit length. Figure 4.7 shows such a mesh with the number of nodes per side, n_s, equal to five. The method of generation can follow fairly closely that described in the last subsection.

The average number of nodal points per horizontal row is $\frac{1}{2}(n_s + 1)$. Since there are n_s such rows the total number of nodes is $\frac{1}{2}n_s(n_s + 1)$. In the first horizontal row of elements at the bottom of the mesh there are $2n_s - 3$ elements, and in the last there is only one. The average number of elements per

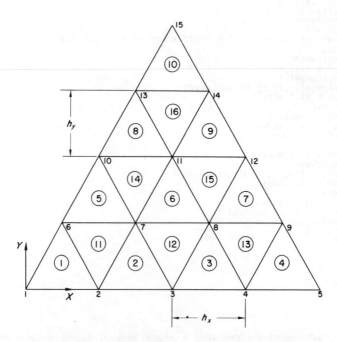

Figure 4.7   A triangular mesh of equilateral elements

row is therefore $n_s - 1$, and since there are $n_s - 1$ such rows the total number of elements is $(n_s - 1)^2$. For unit length of mesh side, the horizontal distances between nodal points on the same row, h_x, and the vertical distances between such rows, h_y, are

$$h_x = 1/(n_s - 1), \quad h_y = \tfrac{1}{2}h_x \sqrt{3} \tag{4.8}$$

Let i_y be used to count horizontal rows of nodes from the bottom to the top of the mesh, where $1 \leqslant i_y \leqslant n_s$. Also, let i_x be used to count nodes along a particular horizontal row from left to right, where $1 \leqslant i_x \leqslant n_x$, n_x being the number of nodes in that row and equal to $n_s - i_y + 1$. Assuming that the origin of the global co-ordinates is at the first node, the co-ordinates of the typical node defined by i_x and i_y are

$$X_i = (i_x - 1)h_x + \tfrac{1}{2}(i_y - 1)h_x, \quad Y_i = (i_y - 1)h_y \tag{4.9}$$

As in the last subsection it is convenient to consider first all the upright elements, such as those numbered 1 to 10 in figure 4.7. Let i_y now be used to count horizontal rows of elements from the bottom to the top of the mesh, where $1 \leqslant i_y \leqslant n_s - 1$. Also, let the bottom left-hand corner of each element be the local origin. For the typical upright element numbered m, the numbers of its nodes may be defined in anticlockwise order as

$$i = m + i_y - 1, \quad j = i + 1, \quad k = m + n_s \tag{4.10}$$

Let the total number of upright elements be m_1.

Inverted elements such as those numbered 11 to 16 in figure 4.7 are then considered. Let i_y again be used to count horizontal rows of elements and let the bottom corner of each element be the local origin. For the typical inverted element numbered m, the numbers of its nodes may be defined as

$$i = m - m_1 + 2i_y - 1, \quad j = m - m_1 + n_s + i_y, \quad k = j - 1 \tag{4.11}$$

Figure 4.8 shows a version of subprogram MESH for generating mesh data by the above method. The variable names are the same as those used in figure 4.6, the only additions being HX and HY which store the mesh dimensions h_x and h_y, and NSPT which stores the value of n_s. The procedure is also very similar, but with the changes required by the above analysis.

4.3.4 A circular mesh Circular meshes can be designed to suit a range of practical problems. The first consideration is the type of element size distribution required. For example, it may be desirable to have a relatively large number of small elements near the centre of the circle. If the elements are arranged in circular rings this could be achieved by, say, using the same number of elements in each ring and progressively reducing the widths of the rings towards the centre. A mesh based on this principle is described in section 7.3. Alternatively, a circular mesh containing elements of approximately uniform size may be required.

```
          SUBROUTINE  MESH
C
C  SUBPROGRAM TO READ OR GENERATE A MESH OF TRIANGULAR FINITE ELEMENTS.
C  THIS VERSION GENERATES A UNIFORM EQUILATERAL TRIANGULAR MESH.
C
          COMMON /CMESH/ NEL,NNP,X(121),Y(121),AI(200),AJ(200),AK(200),
       1      BI(200),BJ(200),BK(200),AREA(200),NPI(200),NPJ(200),NPK(200),
       2      NBP,NPB(40),MOUT
       3      /CMPAR/ NSPT
C
C  INPUT THE NUMBER OF POINTS ON EACH SIDE OF THE MESH. ALSO THE MESH
C  DATA OUTPUT CONTROL PARAMETER.
          READ(5,51) NSPT,MOUT
    51    FORMAT(2I5)
C
C  COMPUTE AND TEST THE NUMBERS OF NODES AND ELEMENTS.
          NNP=NSPT*(NSPT+1)/2
          NEL=(NSPT-1)**2
          IF(NNP.LE.121.AND.NEL.LE.200) GO TO 1
          WRITE(6,61) NNP,NEL
    61    FORMAT(30HOEXCESSIVE SIZE OF MESH, NNP =,I5,8H,  NEL =,I5)
          STOP
C
C  DEFINE THE NODAL POINT CO-ORDINATES.
     1    HX=1./FLOAT(NSPT-1)
          HY=HX*0.5*SQRT(3.)
          I=0
          DO 2 IY=1,NSPT
          NXPT=NSPT-IY+1
          DO 2 IX=1,NXPT
          I=I+1
          X(I)=FLOAT(IX-1)*HX+FLOAT(IY-1)*0.5*HX
     2    Y(I)=FLOAT(IY-1)*HY
C
C  DEFINE THE NUMBERS OF THE THREE NODES OF EACH ELEMENT.
          M=0
          NYEL=NSPT-1
          DO 3 IY=1,NYEL
          NXEL=NSPT-IY
          DO 3 IX=1,NXEL
          M=M+1
          NPI(M)=M+IY-1
          NPJ(M)=NPI(M)+1
     3    NPK(M)=M+NSPT
          M1=M
          NYEL=NYEL-1
          DO 4 IY=1,NYEL
          NXEL=NSPT-IY-1
          DO 4 IX=1,NXEL
          M=M+1
          NPI(M)=M-M1+2*IY-1
          NPJ(M)=M-M1+NSPT+IY
     4    NPK(M)=NPJ(M)-1
          RETURN
          END
```

Figure 4.8   Subprogram for generating a triangular mesh of equilateral elements

Figure 4.9 shows a mesh of this latter type with the number of elements at the centre, n_c, and the number of nodal points along a horizontal radius, n_r, equal to six and four respectively. Normally an integer value between five and eight is assigned to n_c to ensure that element angles at the centre are reasonably close to the equilateral value of $60°$. The value of n_r is the main parameter determining the total number of elements in the mesh. Let i_r be used to count outwards from the centre both the rings of elements and the rings of nodal

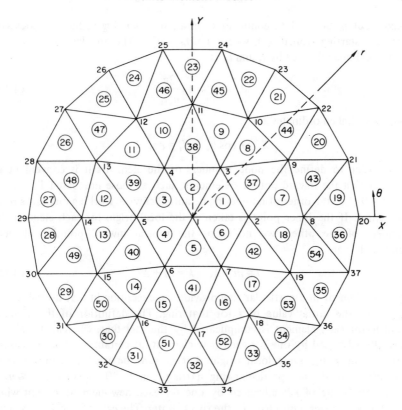

Figure 4.9 A circular mesh

points, ignoring the centre point, where $1 \leqslant i_r \leqslant n_r - 1$. If the mesh is of unit radius and the radial spacing of the rings of nodes is uniform, then the radius of a typical ring defined by i_r is

$$r = i_r/(n_r - 1) \qquad (4.12)$$

The number of points per ring, $n_c i_r$, is proportional to the radius in order to keep the sizes of the elements approximately uniform. Since the innermost and outermost rings contain respectively n_c and $n_c(n_r - 1)$ points, the average number of points per ring is $\frac{1}{2}n_c n_r$ and the total number of nodes is $\frac{1}{2}n_c n_r(n_r - 1) + 1$. The number of elements per ring is $n_c(2i_r - 1)$. Since the innermost and outermost rings contain respectively n_c and $n_c(2n_r - 3)$ elements, the average number per ring is $n_c(n_r - 1)$ and the total number of elements is $n_c(n_r - 1)^2$.

Let the origin of the global co-ordinates X and Y be at the centre of the circle, and in addition to the radial co-ordinate used in equation 4.12 let there be an angular co-ordinate θ measured in the anticlockwise direction from the

X-axis. Let i_θ be used to count nodes in a particular ring in the anticlockwise direction starting from $\theta = 0$, where $1 \leqslant i_\theta \leqslant n_c i_r$. The angular co-ordinate of the typical node defined by i_r and i_θ is

$$\theta = (i_\theta - 1)\frac{2\pi}{n_c i_r} \qquad (4.13)$$

and the global co-ordinates are

$$X_i = r \cos \theta, \quad Y_i = r \sin \theta \qquad (4.14)$$

In order to define the node numbers of the elements it is convenient to consider first the inward pointing elements, such as those numbered 1 to 36 in figure 4.9. Let the order of numbering be anticlockwise, working outwards from the centre. If the inner point is taken as the local origin for each element the numbers of the other two nodes of the typical inward pointing element numbered m are

$$j = m + 1, \quad k = m + 2 \qquad (4.15)$$

The last element in a particular ring provides an exception in that the value of k reverts to the initial value of j for that ring. The definition of the 'i' node number for each element is complicated by the fact that, whereas its value is normally advanced by one for each new element, at certain intervals successive elements have the same local origin. Let i_θ now be used to count inward pointing elements in a particular ring starting from $\theta = 0$, where $1 \leqslant i_\theta \leqslant n_c i_r$. Hence, the value of i is advanced by one for each new element, except when $i_\theta - 1$ is an integer multiple of i_r, the ring counter. The case of $i_\theta = 1$ is a special one representing the first inward pointing element in a particular ring, the local origin of which is the 'k' node of the previous element. Also, for the last element in a ring the value of i reverts to its initial value for the ring. Consider the mesh shown in figure 4.9. In the first ring of elements containing elements 1 to 6, $i_\theta - 1$ is always an integer multiple of one and all six elements have the same local origin. In the second ring containing inward pointing elements 7 to 18, the 'i' node of element 7 is the 'k' node of element 6, that is node 2. For element 8, $i_\theta - 1 = 1$ and the value of i is advanced by one from that for element 7. For element 9, however, $i_\theta - 1 = 2$ which is an integer multiple of two, the ring number, and the value of i remains as for element 8: and so on. Let the total number of inward pointing elements be m_1.

Outward pointing elements such as those numbered 37 to 54 in figure 4.9 are then considered: note that this type start in the second ring. The local origin point for the typical outward pointing element numbered m can be defined as

$$i = m - m_1 + 1 \qquad (4.16)$$

The number of the third node can be defined as $k = i + 1$, except for the last element in a particular ring when the value of k reverts to the initial value of i for that ring. The number of the second node, j, can best be defined by advancing its value by one for each new element, except when $i_\theta - 1$ is an integer multiple of $i_r - 1$ and the value of j is advanced by two. The parameter i_θ now counts outward pointing elements in a particular ring starting from $\theta = 0$, where $1 \leqslant i_\theta \leqslant n_c(i_r - 1)$. The case of $i_\theta = 1$ is again a special one where the value of j is found by adding two to the corresponding value for the previous element. For example, in figure 4.9 the value of j for element 43 is 21, which is two more than the corresponding value for element 42.

Figure 4.10 shows a version of subprogram MESH for generating mesh data by the above method. The variables NCEL, NRPT, ITH and IR are used to store the values of n_c, n_r, i_θ and i_r respectively, while I, M, M1, R, THETA and PI serve to store the values of i, m, m_1, r, θ and π. NREL stores the number of rings of elements and NTHPT stores the numbers of nodal points or elements in a particular ring. The variable J stores the value of j for the first outward pointing element in a new ring, while K stores the value of k for the last inward pointing element in a particular ring. In the context of element node numbering, the variable I stores the value of i for the first and last inward pointing elements in a particular ring.

```
      SUBROUTINE   MESH
C
C   SUBPROGRAM TO READ OR GENERATE A MESH OF TRIANGULAR FINITE ELEMENTS.
C   THIS VERSION GENERATES A CIRCULAR MESH.
C
      COMMON /CMESH/ NEL,NNP,X(121),Y(121),AI(200),AJ(200),AK(200),
     1     BI(200),BJ(200),BK(200),AREA(200),NPI(200),NPJ(200),NPK(200),
     2     NBP,NPB(40),MOUT
     3          /CMPAR/ NCEL,NRPT
C
C   INPUT THE NUMBER OF ELEMENTS AT CENTRE AND POINTS ALONG A RADIUS,
C   ALSO THE MESH DATA OUTPUT CONTROL PARAMETER.
      READ(5,51) NCEL,NRPT,MOUT
   51 FORMAT(3I5)
C
C   COMPUTE AND TEST THE NUMBERS OF NODES AND ELEMENTS.
      NNP=NCEL*NRPT*(NRPT-1)/2+1
      NEL=NCEL*(NRPT-1)**2
      IF(NNP.LE.121.AND.NEL.LE.200) GO TO 1
      WRITE(6,61) NNP,NEL
   61 FORMAT(30HOEXCESSIVE SIZE OF MESH, NNP =,I5,8H,   NEL =,I5)
      STOP
C
C   DEFINE THE NODAL POINT CO-ORDINATES.
    1 X(1)=0.
      Y(1)=0.
      PI=4.*ATAN(1.)
      I=1
      NREL=NRPT-1
      DO 2 IR=1,NREL
      R=FLOAT(IR)/FLOAT(NREL)
      NTHPT=NCEL*IR
      DO 2 ITH=1,NTHPT
      THETA=FLOAT(ITH-1)*2.*PI/FLOAT(NTHPT)
      I=I+1
      X(I)=R*COS(THETA)
    2 Y(I)=R*SIN(THETA)
```

Figure 4.10   Subprogram for generating a circular mesh

```
C
C DEFINE THE NUMBERS OF THE THREE NODES OF EACH ELEMENT.
      M=0
      I=1
      DO 4 IR=1,NREL
      NTHPT=NCEL*IR
      DO 3 ITH=I,NTHPT
      M=M+1
      IF(ITH.EQ.1) NPI(M)=I
      IF(ITH.GT.1) NPI(M)=NPI(M-1)+1
      IF(ITH.GT.1.AND.MOD(ITH-1,IR).EQ.0) NPI(M)=NPI(M-1)
      NPJ(M)=M+1
      NPK(M)=M+2
      IF(ITH.GT.1) GO TO 3
      I=NPI(M)
      K=NPJ(M)
    3 CONTINUE
      NPI(M)=I
      NPK(M)=K
    4 I=K
      M1=M
      J=NCEL+3
      DO 6 IR=2,NREL
      NTHPT=NCEL*(IR-1)
      DO 5 ITH=1,NTHPT
      M=M+1
      NPI(M)=M-M1+1
      IF(ITH.EQ.1) NPJ(M)=J
      IF(ITH.GT.1) NPJ(M)=NPJ(M-1)+1
      IF(ITH.GT.1.AND.MOD(ITH-1,IR-1).EQ.0) NPJ(M)=NPJ(M-1)+2
      NPK(M)=NPI(M)+1
    5 IF(ITH.EQ.1) K=NPI(M)
      NPK(M)=K
    6 J=NPJ(M)+2
      RETURN
      END
```

Figure 4.10   Continued

4.4 Mesh Modification

The purpose of mesh modification is to adapt a basic form of mesh to suit the requirements of a particular problem. The first consideration is to fit the shape of the solution domain boundary, but it may also be important to have a particular type of distribution of elements within the domain as discussed in section 4.1. In the finite element computer programs shown in figures 3.10 and 6.5, mesh modification is carried out in subprogram MODIFY immediately after the mesh data have been read in or generated in MESH. Note that the calculation of element geometries is not performed until control of execution returns to the main program. Therefore, the co-ordinates of the nodal points can be changed in MODIFY, usually in such a way as to avoid changing the relative positions of the nodes. Extra elements and nodes can also be added to the basic mesh.

A very simple example of mesh modification is provided by the formation of a rectangular mesh from a square one. Suppose the square mesh with sides of unit length shown in figure 4.11a is to be modified into a rectangle of width W and depth H as shown in figure 4.11b. The data for the basic square mesh might have been either read in or generated by one of the methods described in

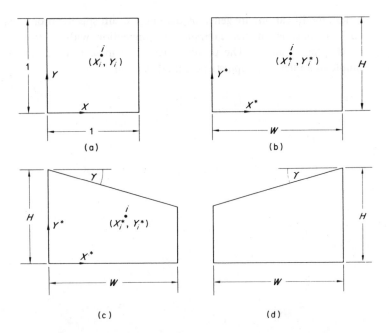

Figure 4.11   Mesh modifications: (a) outline of basic square mesh; (b) rectangular mesh;
(c) and (d) tapered rectangular meshes

sections 4.3.1 and 4.3.2. The modification can be achieved by scaling the global
co-ordinates of the nodal points. For the typical node i shown in figures 4.11a
and 4.11b

$$X_i^* = WX_i, \quad Y_i^* = HY_i \tag{4.17}$$

where the asterisks indicate modified values. Uniformity of element size is
retained. Similar linear scaling can be applied to the triangular and circular
meshes shown in figures 4.7 and 4.9 to make them respectively isosceles and
elliptical in shape.

The modification principle can be extended to produce more complicated
shapes such as the tapered one shown in figure 4.11c from the square one shown
in figure 4.11a by redefining the co-ordinates as

$$X_i^* = WX_i, \quad Y_i^* = (H - X_i^* \tan \gamma)Y_i \tag{4.18}$$

where γ is the angle of taper. Note that, if a mesh of right-angled triangles of the
form shown in figure 4.2 is used for this purpose, the elements do not become
obtuse-angled. If a similar modification is applied to produce the shape shown in
figure 4.11d, however, obtuse-angled elements result, and a different basic mesh
configuration should be employed.

Since mesh modifications are generally designed to suit specific problems, it is convenient to consider further examples in connection with particular case studies in chapters 5 and 7. The problem described in section 7.3 provides a good example of a relatively sophisticated modification.

5 Some Harmonic Problems

The case studies described in this chapter provide practical examples of the application of the finite element analysis and computer program described in chapter 3 to problems of the harmonic type outlined in chapter 2. The problems selected are mostly relatively trivial in that they are amenable to analytical solution. Nevertheless, it is essential to have such test cases against which to compare the results produced by any finite element analysis and program. More complicated problems can then be approached with a reasonable degree of confidence in the method.

5.1 Case Study: Downstream Viscous Flow in a Rectangular Channel

The finite element analysis described in chapter 3 is developed with the aid of a direct equilibrium formulation applied to the downstream viscous flow problem outlined in section 2.2.1. Such a problem is therefore an appropriate choice for this first case study. If the cross-section of the uniform channel is rectangular in shape as shown in figure 2.2, analytical solutions for the velocity profile and volumetric flow rate can be obtained. Also, some results obtained by a finite difference method for a problem of this type are presented by Fenner (1974) and provide interesting comparisons with the present finite element results.

5.1.1 Analytical solution If the rectangular channel is infinitely wide compared with its depth, that is $H \ll W$ in figure 2.2, the governing differential equation 2.31 for the downstream flow is reduced to

$$\frac{d^2 w}{dy^2} = \frac{P_z}{\mu} \qquad (5.1)$$

Integration of this ordinary differential equation with the boundary conditions defined in equations 2.34 yields the following velocity profile

$$w = \frac{y V_z}{H} + \frac{P_z}{2\mu}(y^2 - yH) \qquad (5.2)$$

The volumetric downstream flow rate can be found with the aid of equation 2.35 as

$$Q = \frac{WHV_z}{2} - \frac{WP_z H^3}{12\mu} \qquad (5.3)$$

For a channel of finite width the flow rate can be expressed in a similar form as

$$Q = \frac{WHV_z}{2} F_D - \frac{WP_z H^3}{12\mu} F_P \qquad (5.4)$$

The parameters F_D and F_P are known as the drag and pressure flow shape factors respectively, because of their associations with the drag flow induced by the boundary velocity V_z and the pressure flow due to the gradient P_z. The values of these factors for a particular value of the ratio of channel depth to width H/W may be obtained from

$$F_D = 16 \frac{W}{H} \sum_{i=1,3,5,\ldots}^{\infty} \frac{\tanh(\pi i H/2W)}{\pi^3 i^3} \qquad (5.5)$$

$$F_P = 1 - 192 \frac{H}{W} \sum_{i=1,3,5,\ldots}^{\infty} \frac{\tanh(\pi i W/2H)}{\pi^5 i^5} \qquad (5.6)$$

Fenner (1974) describes a computer program for evaluating pressure flow shape factors, together with a finite difference analysis and results for pressure flow when $H/W = 0.5$.

5.1.2 Problem specification

The pressure flow shape factor is to be computed by the finite element method, for a channel depth to width ratio of $H/W = 0.5$. Uniform meshes of right-angled triangular elements of the basic form described in section 4.3.1 are to be used. Solutions are to be obtained for meshes involving 6×6 and 11×11 nodal points in the solution domain, and compared with both finite difference results and the analytical solution which gives $F_P = 0.6860$ to four significant figures. The effect of taking advantage of the symmetry of the problem to reduce the size of the domain analysed is also to be examined.

Using the same basic form of mesh, flow rate results are to be computed for combined drag and pressure flow and a channel depth to width ratio of $H/W = 0.2$. In this case the effect of modifying the distribution of the nodal points within the solution domain is to be examined with a view to improving the accuracy of the solutions without increasing the number of nodes.

5.1.3 Finite element solution for pressure flow

A computer program for solving this problem is described in section 3.8. Figure 3.10 shows the main program and figures 3.11 and 3.12 show subprograms MSHOUT and SOLVE1 for writing out the mesh data and solving the overall linear equations. The subprogram MESH shown in figure 4.4 is used to generate data for a square mesh

```
      SUBROUTINE  MODIFY
C
C  SUBPROGRAM TO MODIFY THE MESH.
C  THIS VERSION APPLIES LINEAR SCALING TO THE NODE CO-ORDINATES.
C
      COMMON /CMESH/ NEL,NNP,X(121),Y(121),AI(200),AJ(200),AK(200),
     1    BI(200),BJ(200),BK(200),AREA(200),NPI(200),NPJ(200),NPK(200),
     2    NBP,NPB(40),MOUT
C
C  INPUT THE DEPTH (Y-DIRECTION) AND WIDTH (X-DIRECTION).
      READ(5,51) H,W
   51 FORMAT(2F10.0)
C
C  MODIFY THE CO-ORDINATES OF THE NODAL POINTS.
      DO 1 I=1,NNP
      X(I)=X(I)*W
    1 Y(I)=Y(I)*H
      RETURN
      END
```

Figure 5.1   Subprogram for modifying a square mesh to a rectangular form

of right-angled triangular elements. Since the sides of this basic mesh are of unit length, the node co-ordinate modifications defined by equations 4.17 are applied to form the required rectangular mesh. Figure 5.1 shows an appropriate version of subprogram MODIFY for this purpose, the required values of channel depth and width being read in. The other three subprograms required are PHI1F to define the function ϕ_1 appearing in the general governing differential equation 2.87, BCS to apply the boundary conditions, and OUTPUT to write out the results.

In the present example $\phi_1 = P_z/\mu$, and figure 5.2 shows a suitable version of PHI1F for reading the values of P_z and μ into variables PZ and VISCOS respectively, writing them out and then defining ϕ_1 for every element in the mesh. The boundary conditions for pressure flow are zero prescribed values of velocity at every boundary node. Figure 5.3 shows a suitable version of BCS for reading the boundary node numbers, writing them out, and then modifying the overall equations according to equations 3.62. More sophisticated versions of BCS could be used to apply nonzero values of the unknowns and to generate the boundary node numbers from the mesh parameters (in this case NXPT and NYPT used in MESH).

```
      SUBROUTINE  PHI1F(PHI1,NEL)
C
C  SUBPROGRAM TO DEFINE THE MEAN VALUE OF THE PHI1 FUNCTION IN THE
C  HARMONIC DIFFERENTIAL EQUATION FOR EACH ELEMENT IN THE MESH.
C
      DIMENSION  PHI1(200)
      READ(5,51) PZ,VISCOS
   51 FORMAT(2E15.5)
      WRITE(6,61) PZ,VISCOS
   61 FORMAT(20HOPRESSURE GRADIENT =,E12.4,10X,11HVISCOSITY =,E12.4)
      DO 1 M=1,NEL
    1 PHI1(M)=PZ/VISCOS
      RETURN
      END
```

Figure 5.2   Subprogram for defining the distribution of function ϕ_1 for downstream viscous flow

```
      SUBROUTINE  BCS
C
C  SUBPROGRAM TO APPLY THE BOUNDARY CONDITIONS.
C  THIS VERSION PRESCRIBES ZERO VALUES OF THE UNKNOWNS.
C
      COMMON /CMESH/ NEL,NNP,X(121),Y(121),AI(200),AJ(200),AK(200),
     1      BI(200),BJ(200),BK(200),AREA(200),NPI(200),NPJ(200),NPK(200),
     2      NBP,NPB(40),MOUT
     3             /CEQNS/ OSTIFF(121,9),DELTA(121),F(121),NPA(121,9),NAP(121)
C
C  INPUT THE BOUNDARY NODE NUMBERS.
      READ(5,51) NBP
  51  FORMAT(14I5)
      IF(NBP.LE.40) GO TO 1
      WRITE(6,61) NBP
  61  FORMAT(43HOEXCESSIVE NUMBER OF BOUNDARY POINTS, NBP =,I5)
      STOP
   1  READ(5,51) (NPB(I),I=1,NBP)
C
C  APPLY ZERO VALUES OF THE UNKNOWNS AT THE BOUNDARY POINTS.
      FACT=1.E10
      DO 2 I=1,NBP
      IROW=NPB(I)
      OSTIFF(IROW,1)=OSTIFF(IROW,1)*FACT
   2  F(IROW)=0.
C
C  OUTPUT THE BOUNDARY POINT NUMBERS.
      WRITE(6,62) NBP,(NPB(IB),IB=1,NBP)
  62  FORMAT(20HOTHE NUMBERS OF THE ,I3,2OH BOUNDARY POINTS ARE /
     1          (14(1X,I4)))
      RETURN
      END
```

Figure 5.3   Subprogram for applying the boundary conditions

The result required in the present problem is the volumetric flow rate. Since the velocity distributions over CST elements are linear, the integration defined by equation 2.35 can be achieved by the following summation for all the elements

$$Q = \sum \frac{\Delta m}{3} (w_i + w_j + w_k) \qquad (5.7)$$

Figure 5.4 shows a version of OUTPUT for carrying out this computation and writing out the value of the integral. The values of the nodal point velocities stored in array DELTA could also be written out.

According to equation 5.4 the pressure flow shape factor for $H/W = 0.5$ is equal to the computed flow rate when the following data are used: $H = 1$, $W = 2$, $\mu = 1$, $P_z = -6$, in addition to the zero value of V_z already applied in BCS. Figure 5.5 reproduces the corresponding input data required for a mesh involving 6 x 6 nodal points. The items of data may be listed as follows.

(1) Problem title required by the main program (figure 3.10).

(2) The numbers of nodal points in the co-ordinate directions and a nonzero value of the mesh data output control parameter read into MESH (figure 4.4).

(3) The channel depth and width read into MODIFY (figure 5.1).

```
      SUBROUTINE   OUTPUT
C
C  SUBPROGRAM TO OUTPUT THE FINAL RESULTS.
C
      COMMON /CMESH/ NEL,NNP,X(121),Y(121),AI(200),AJ(200),AK(200),
     1       BI(200),BJ(200),BK(200),AREA(200),NPI(200),NPJ(200),NPK(200),
     2       NBP,NPB(40),MOUT
     3          /CEQNS/ OSTIFF(121,9),DELTA(121),F(121),NPA(121,9),NAP(121)
C
C  INTEGRATE OVER THE SOLUTION DOMAIN.
      SUM=0.
      DO 1 M=1,NEL
      I=NPI(M)
      J=NPJ(M)
      K=NPK(M)
      DMEAN=(DELTA(I)+DELTA(J)+DELTA(K))/3.
    1 SUM=SUM+DMEAN*AREA(M)
      WRITE(6,61) SUM
   61 FORMAT(36HOINTEGRAL OVER THE SOLUTION DOMAIN =,E13.5)
      RETURN
      END
```

Figure 5.4   Subprogram for integrating the dependent variable over the solution domain
and writing out the results

(4) The values of P_z and μ read into PHI1F (figure 5.2).

(5) The total number of boundary nodes and the numbers assigned to them, required by BCS (figure 5.3).

(6) The maximum number of cycles of iteration, output frequency, over-relaxation factor and convergence tolerance required by SOLVE1 (figure 3.12).

Figure 5.6 shows the printed results produced by the program.

These results are for the finite element analysis applied to the entire channel cross-section with the boundary conditions as shown in figure 5.7a. Since pressure flow is symmetrical about the two dotted lines shown in this diagram, it is possible to consider only one quadrant of the solution domain. For example, figure 5.7b shows the shaded quadrant with the appropriate dimensions and boundary conditions that take account of symmetry. The two cases can be distinguished by defining a number of boundary condition (NBC) parameter as shown. The case NBC = 2 can be accommodated by reading values of 0.5 and 1.0 into H and W in subprogram MODIFY, and only reading the numbers of the nodes on the top and right-hand side of the mesh into subprogram BCS. As

```
SLOW VISCOUS FLOW ALONG A RECTANGULAR CHANNEL
    6    6    1
 1.        2.
   -6.00000E+00    1.00000E+00
   20
    1    2    3    4    5    6    7   12   13   18   19   24   25   30
   31   32   33   34   35   36
  300    3 1.35      0.000001
```

Figure 5.5   Typical input data for the harmonic program applied to the downstream
viscous flow problem

CST FINITE ELEMENT SOLUTION FOR TWO-DIMENSIONAL HARMONIC PROBLEM

SLOW VISCOUS FLOW ALONG A RECTANGULAR CHANNEL

PRESSURE GRADIENT = -0.6000E+01   VISCOSITY = 0.1000E+01

GEOMETRIC DATA FOR THE MESH

NUMBER OF ELEMENTS = 50

NUMBER OF NODAL POINTS = 36

NODAL POINT CO-ORDINATES

I	X	Y	I	X	Y	I	X	Y
1	0.0000	0.0000	2	0.4000	0.0000	3	0.8000	0.0000
4	1.2000	0.0000	5	1.6000	0.0000	6	2.0000	0.0000
7	0.0000	0.2000	8	0.4000	0.2000	9	0.8000	0.2000
10	1.2000	0.2000	11	1.6000	0.2000	12	2.0000	0.2000
13	0.0000	0.4000	14	0.4000	0.4000	15	0.8000	0.4000
16	1.2000	0.4000	17	1.6000	0.4000	18	2.0000	0.4000
19	0.0000	0.6000	20	0.4000	0.6000	21	0.8000	0.6000
22	1.2000	0.6000	23	1.6000	0.6000	24	2.0000	0.6000
25	0.0000	0.8000	26	0.4000	0.8000	27	0.8000	0.8000
28	1.2000	0.8000	29	1.6000	0.8000	30	2.0000	0.8000
31	0.0000	1.0000	32	0.4000	1.0000	33	0.8000	1.0000
34	1.2000	1.0000	35	1.6000	1.0000	36	2.0000	1.0000

ELEMENT NODE NUMBERS AND AREAS

M	I	J	K	AREA	M	I	J	K	AREA
1	1	8	7	0.4000E-01	2	1	2	8	0.4000E-01
3	2	9	8	0.4000E-01	4	2	3	9	0.4000E-01
5	3	10	9	0.4000E-01	6	3	4	10	0.4000E-01
7	4	11	10	0.4000E-01	8	4	5	11	0.4000E-01
9	5	12	11	0.4000E-01	10	5	6	12	0.4000E-01
11	7	14	13	0.4000E-01	12	7	8	14	0.4000E-01
13	8	15	14	0.4000E-01	14	8	9	15	0.4000E-01
15	9	16	15	0.4000E-01	16	9	10	16	0.4000E-01
17	10	17	16	0.4000E-01	18	10	11	17	0.4000E-01
19	11	18	17	0.4000E-01	20	11	12	18	0.4000E-01
21	13	20	19	0.4000E-01	22	13	14	20	0.4000E-01
23	14	21	20	0.4000E-01	24	14	15	21	0.4000E-01
25	15	22	21	0.4000E-01	26	15	16	22	0.4000E-01
27	16	23	22	0.4000E-01	28	16	17	23	0.4000E-01
29	17	24	23	0.4000E-01	30	17	18	24	0.4000E-01
31	19	26	25	0.4000E-01	32	19	20	26	0.4000E-01
33	20	27	26	0.4000E-01	34	20	21	27	0.4000E-01
35	21	28	27	0.4000E-01	36	21	22	28	0.4000E-01
37	22	29	28	0.4000E-01	38	22	23	29	0.4000E-01
39	23	30	29	0.4000E-01	40	23	24	30	0.4000E-01
41	25	32	31	0.4000E-01	42	25	26	32	0.4000E-01
43	26	33	32	0.4000E-01	44	26	27	33	0.4000E-01
45	27	34	33	0.4000E-01	46	27	28	34	0.4000E-01
47	28	35	34	0.4000E-01	48	28	29	35	0.4000E-01
49	29	36	35	0.4000E-01	50	29	30	36	0.4000E-01

THE NUMBERS OF THE 20 BOUNDARY POINTS ARE

1	2	3	4	5	6	7	12	13	18	19	24	25	30
31	32	33	34	35	36								

SOLUTION OF EQUATIONS BY GAUSS-SEIDEL ITERATION

OVER-RELAXATION FACTOR = 1.350

ITER	ERROR
3	0.1132E+00
6	0.4074E-02
9	0.2743E-03
12	0.1740E-04

ITERATION CONVERGED TO A TOLERANCE OF 1.0000E-06 AFTER   15 CYCLES

INTEGRAL OVER THE SOLUTION DOMAIN = 0.59802E+00

Figure 5.6   Printed results obtained from the harmonic program supplied with input data shown in figure 5.5

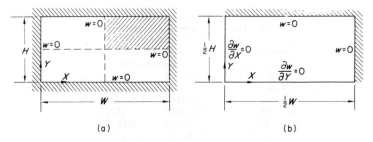

Figure 5.7   Alternative solution domains and boundary conditions for downstream viscous flow problem: (a) number of boundary condition (NBC) = 1; (b) NBC = 2

indicated in section 3.5, zero normal derivative boundary conditions are obtained by treating the relevant boundary nodes as internal points. Since the resulting flow rate integral is for one quadrant of the domain, its value must be multiplied by four to give the shape factor.

5.1.4 Results for pressure flow   Before studying the computed results for shape factor, it is necessary to establish both the optimum over-relaxation factors, and the appropriate tolerance levels. The former are determined empirically by finding the numbers of cycles of iteration required for convergence to a tolerance of 10^{-2}, for values of ω in the range $1 < \omega < 2$. Figure 5.8 shows the results for 11 x 11 point (121 node, 200 element) meshes. The shape of the curves for both types of boundary condition is typical of the considerable effect of over-relaxation. The number of cycles decreases steadily

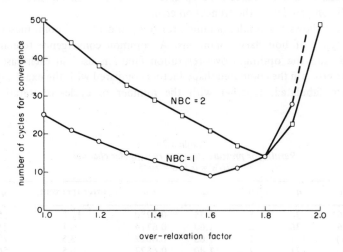

Figure 5.8   Effect of over-relaxation on convergence to a tolerance of 10^{-2}, for 11 x 11 point meshes

Table 5.1
Effect of convergence tolerance for
11 × 11 point meshes

| Convergence | Shape factors | |
tolerance	NBC = 1	NBC = 2
10^{-2}	0.66367	0.69540
10^{-3}	0.66330	0.68212
10^{-4}	0.66216	0.68192
10^{-5}	0.66221	0.68224
10^{-6}	0.66222	0.68220

to a minimum as ω is increased, and then increases rather more rapidly (for NBC = 1 and $\omega = 2$, the process failed to converge in 500 cycles). If the optimum ω is not known, it is safer to underestimate rather than to overestimate its value. This optimum value is affected both by the type of boundary conditions and by the number of nodal points.

While a tolerance of 10^{-2} is appropriate for examining the effect of over-relaxation, it is too large to give accurate results for shape factor. Table 5.1 shows the variation of the computed value of shape factor with the convergence tolerance, using 11 × 11 point meshes and the relevant optimum over-relaxation factors. Two sources of error are involved: the truncation errors in the finite element method and the convergence errors in the Gauss—Seidel process. If the latter are negligible, the value of shape factor should be independent of the convergence tolerance. Any difference between the computed value and the analytical solution is then due to truncation errors. Clearly, the convergence errors still have a small effect when the tolerance is as small as 10^{-6}. Nevertheless, such a tolerance is acceptable because the convergence errors are then small compared with the truncation errors.

Table 5.2 shows the results obtained for 6 × 6 and 11 × 11 point meshes and the two types of boundary conditions. A common convergence tolerance is employed, and the optimum over-relaxation factor is used in each case. The percentage errors in the computed shape factors compared with the exact value of 0.6860 are tabulated, together with the number of cycles of iteration, q,

Table 5.2
Results for pressure flow in a rectangular channel:
$H/W = 0.5$, $\alpha = 10^{-6}$

Mesh	n	NBC	ω	F_P	Error (per cent)	q
6 × 6	36	1	1.35	0.5980	12.8	15
6 × 6	36	2	1.60	0.6714	2.1	26
11 × 11	121	1	1.60	0.6622	3.5	28
11 × 11	121	2	1.80	0.6822	0.5	56

required for convergence. Although the value of q for NBC = 2 is about twice that for NBC = 1 using the same mesh, the computed shape factor is much more accurate. In fact, the result obtained using a 6 x 6 point mesh and NBC = 2 is more accurate than that for an 11 x 11 point mesh and NBC = 1, and requires slightly fewer iterations. Since the number of nodal points, n, and hence the number of linear equations to be solved, differ by a factor of nearly four, it is clear that taking advantage of the symmetry of the problem results in a saving of computer time and storage requirement of about 75 per cent.

It is interesting to compare the present results with those obtained by the finite difference method described in detail by Fenner (1974) and outlined here in section 2.4.1. For NBC = 1 all the results shown in figure 5.8 and tables 5.1 and 5.2 are identical in every detail with the equivalent finite difference ones for the same number of points in the solution domain. This confirms the close similarity between the methods demonstrated by the comparison described in section 3.3. On the other hand, there are considerable differences for NBC = 2: the finite element results are substantially more accurate and are obtained with less computation than the equivalent finite difference ones. This is due to the different methods of applying derivative boundary conditions.

In section 3.7.3 it is predicted that the present finite element method provides lower bound solutions for velocities, and hence for volumetric flow rates. This prediction is proved correct for the present problem since the computed shape factors are all lower than the true value.

There are various ways to improve the accuracy of the present results. The most obvious is to use more nodal points and elements in the mesh: according to the arguments outlined in section 3.7 the method is convergent so that refining the mesh makes the computed solution agree more closely with the true one. The disadvantage is that the cost of obtaining the solution in terms of both computing time and storage increases rapidly as the mesh is refined. A second method is to modify the distribution of elements within the mesh to improve accuracy without increasing the cost: this approach is used in the next subsection. A third method takes advantage of the known form of the truncation error involved in the finite element method. According to equation 3.79 this error is proportional to the square of the dimensions of the elements. If solutions $w_i^{(1)}$ are obtained when the typical element dimension is h, and another set $w_i^{(2)}$ when it is reduced to $\frac{1}{2}h$ (that is, when four times as many elements are used), then if w_i are the true solutions

$$w_i \approx w_i^{(1)} + \epsilon h^2 \approx w_i^{(2)} + \frac{1}{4}\epsilon h^2 \tag{5.8}$$

where ϵ is the constant of proportionality in the error term. Eliminating this term

$$w_i \approx \frac{1}{3}(4\, w_i^{(2)} - w_i^{(1)}) \tag{5.9}$$

This process of obtaining improved solutions is often referred to as 'h^2 extrapolation', and in the present problem may also be applied to shape factors obtained by integrating velocity profiles. The results shown in table 5.2 for NBC = 2 and meshes of 6 x 6 and 11 x 11 points are respectively 2.1 and 0.5 per cent in error. Using equation 5.9, an improved estimate can be obtained as

$$F_P = \tfrac{1}{3}(4 \times 0.6822 - 0.6714) = 0.6858$$

which is only 0.03 per cent in error.

The relative advantages and disadvantages of direct elimination and iterative methods for solving simultaneous linear equations are discussed in section 3.6.3. In particular, criteria for the Gauss–Seidel method to be faster than the elimination methods are expressed in equations 3.75 and 3.76. These tests may be applied to the results shown in table 5.2, and table 5.3 shows the values obtained for the relative efficiency parameters r_1 and r_2. Since these values are respectively less than and greater than 1 for all the cases considered, the Gauss–Seidel method is faster than the full elimination method but slower than elimination applied to the rectangular form of stiffness matrix. It should be noted, however, that increasing the number of nodes tends to favour the Gauss–Seidel method. This, combined with the relative storage requirements discussed in section 3.6.3, means that the choice between direct and iterative methods is not clearcut, at least for harmonic problems.

Table 5.3
*Relative efficiency parameters for iterative and
direct methods of solution*

Mesh	n	NBC	q	$r_1 = \dfrac{27q}{n^2}$	$r_2 = \dfrac{9q}{n}$
6 x 6	36	1	15	0.31	3.75
6 x 6	36	2	26	0.54	6.50
11 x 11	121	1	28	0.05	2.08
11 x 11	121	2	56	0.10	4.16

5.1.5 Solution and results for combined drag and pressure flow   The second part of the problem specified in section 5.1.2 can be solved with the same set of subprograms as before, except that BCS must be modified to apply nonzero values of velocity along the top boundary of the mesh. The flow is now only symmetrical about the vertical centre line of the channel: figure 5.9 shows the new solution domain and boundary conditions.

The volumetric flow rate is now determined by the values of P_z and V_z, and equation 5.4 can be expressed in the more convenient dimensionless form

$$\pi_Q = \frac{F_D}{2} - \frac{\pi_P F_P}{12} \qquad (5.10)$$

where the dimensionless flow rate and pressure gradient are defined as

$$\pi_Q = \frac{Q}{WHV_z}, \qquad \pi_P = \frac{P_z H^2}{\mu V_z} \qquad (5.11)$$

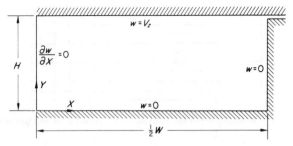

Figure 5.9   Solution domain for analysis of combined drag and pressure flow

For $H/W = 0.2$ the analytical solution defined by equations 5.5 and 5.6 becomes

$$\pi_Q = 0.4457 - 0.07283\pi_P \qquad (5.12)$$

Some results obtained using this equation for particular values of π_P are shown in the second column of table 5.4.

The results shown in the third column of table 5.4 are for a finite element mesh of 11 x 6 nodal points (11 points in the vertical direction and 6 in the horizontal, that is, half of an 11 x 11 point mesh covering the whole channel cross-section) in which all the right-angled triangular elements are of the same size. In a relatively shallow channel significant variations of velocity in the horizontal direction only occur near the side walls. Using the principle outlined in section 4.1 of concentrating relatively small elements in regions of the solution domain where the unknowns change most rapidly, the mesh should be modified to concentrate elements towards the sides of the channel. For the present mesh covering the domain shown in figure 5.9 the following modification to the X co-ordinates (applied in subprogram MODIFY) gives excellent results

$$\frac{X_i^*}{W} = \left(\frac{X_i}{W}\right)^{H/W} \qquad (5.13)$$

Table 5.4
*Dimensionless flow rates for combined drag and
pressure flow for H/W = 0.2, using an 11 x 6 mesh
covering half the channel*

| | | Computed values of π_Q | |
π_P	True π_Q	Uniform mesh	Modified mesh
-2	0.5914	0.5828	0.5890
0	0.4457	0.4403	0.4452
2	0.3000	0.2978	0.3015
4	0.1544	0.1553	0.1577

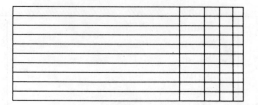

Figure 5.10   Modified mesh for analysis of combined drag and pressure flow

Figure 5.10 shows the horizontal and vertical lines of nodal points in the modified mesh drawn to scale. The dimensionless flow rates obtained using this mesh are shown in the last column of table 5.4. There is a considerable improvement over the results obtained using a uniform mesh, particularly for small π_P. As π_P is increased the conflicting effects of drag and pressure flows tend to spoil the comparison. Note that the use of the long thin elements implied by figure 5.10 causes no difficulties in this harmonic problem.

5.2   Case Study: Torsion of Prismatic Bars

The problem of torsion of a prismatic bar outlined in section 2.2.2 provides a good example of the application of the finite element analysis developed in chapter 3 for downstream fluid flow to other problems of the harmonic type. Bars of any cross-sectional shape can be analysed, but for present purposes elliptical and triangular shapes are considered.

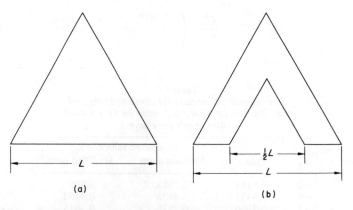

Figure 5.11   Cross-sections of bars in torsion: (a) equilateral triangular section; (b) modified triangular section

5.2.1 Problem specification The torsional stiffness of a bar of elliptical cross-section is to be found using meshes of the basic form described in section 4.3.4. Results are to be obtained for ratios of the semi-axes a_x/a_y (figure 2.3) of 1, 2 and 4, and compared with the analytical solution for this problem. A similar comparison is to be made for the bar of equilateral triangular cross-section shown in figure 5.11a, using meshes of the form described in section 4.3.3. Finally, the effect on torsional stiffness of modifying this triangular section as shown in figure 5.11b is to be determined.

5.2.2 Analytical solutions The governing differential equation for torsion in terms of stress function, χ, is equation 2.45, and the boundary condition is $\chi = 0$. Since the equation defining the boundary of the elliptical section shown in figure 2.3 is

$$\frac{x^2}{a_x^2} + \frac{y^2}{a_y^2} = 1 \tag{5.14}$$

the following choice of stress function automatically satisfies the boundary condition

$$\chi = A \left(\frac{x^2}{a_x^2} + \frac{y^2}{a_y^2} - 1 \right) \tag{5.15}$$

where A is a constant. It also satisfies the differential equation if

$$\nabla^2 \chi = 2A \left(\frac{1}{a_x^2} + \frac{1}{a_y^2} \right) = -2G\theta \tag{5.16}$$

that is, if

$$A = -\frac{G\theta a_x^2 a_y^2}{a_x^2 + a_y^2} \tag{5.17}$$

where θ is the angle of rotation per unit length of the bar. The required torsional couple is given by equation 2.47, which in this case leads to the following result for torsional stiffness

$$\frac{C}{\theta} = \frac{\pi G a_x^3 a_y^3}{a_x^2 + a_y^2} \tag{5.18}$$

The torsional stiffness of the triangular section shown in figure 5.11a may be similarly obtained as

$$\frac{C}{\theta} = \frac{GL^4 \sqrt{3}}{80} \tag{5.19}$$

5.2.3 Finite element solutions and results The torsional stiffnesses of elliptical bars may be found using the set of finite element subprograms described in section 5.1.3, with the exceptions that the version of MESH shown in figure 4.10 is employed, and PHI1F shown in figure 5.2 is modified to define $\phi_1 = -2G\theta = -2$, say, for all the elements. Subprogram MODIFY shown in figure 5.1 can be used to apply the linear scaling necessary to make the basic circular mesh elliptical. A value of one may be read into the variable H to serve as a_y, together with the required value of a_x into W. Subprogram OUTPUT shown in figure 5.4 now computes the integral of the stress function over the solution domain: according to equation 2.47 this value must be doubled to obtain the torsional couple.

Table 5.5 shows the computed results, together with the true values of torsional stiffness derived from equation 5.18. In every case a mesh having six elements at the centre as shown in figure 4.9 is used. For each shape of bar, meshes having 3, 4 and 6 nodal points along a horizontal radius are used. As is to be expected, the accuracies of the computed stiffnesses increase as the meshes are refined. It should be noted, however, that the present mesh modifications which result in relatively long thin elements for large values of a_x/a_y have no effect on the accuracy of the computed stiffness (or on the convergence of the Gauss–Seidel method) for a particular basic mesh.

Table 5.5
Torsional stiffnesses of elliptical bars: $a_y = 1$, $G = 1$, $n_c = 6$

a_x	n_r	Number of nodes	Number of elements	Torsional stiffness True	Torsional stiffness Computed	Error (per cent)
1	3	19	24		1.353	14
	4	37	54	1.571	1.469	6.5
	6	91	150		1.533	2.4
2	3	19	24		4.330	14
	4	37	54	5.026	4.701	6.5
	6	91	150		4.906	2.4
4	3	19	24		10.19	14
	4	37	54	11.83	11.06	6.5
	6	91	150		11.54	2.4

The torsional stiffness of the triangular cross-section shown in figure 5.11a can be found by substituting the version of MESH shown in figure 4.8. Values of one can be read into the scale factors used in MODIFY, to make $L = 1$. Table 5.6 shows the computed results, together with the true value of torsional stiffness derived from equation 5.19. Three meshes are used, having 5, 9 and 13 nodes per side (the case $n_s = 5$ is shown in figure 4.7). The accuracies of both these and the results shown in table 5.5 could be further improved with the aid of the h^2 extrapolation process described in section 5.1.4.

The computed torsional stiffnesses shown in tables 5.5 and 5.6 are all lower than the corresponding true values. At first sight this would appear to violate the

Table 5.6
*Torsional stiffness of a bar of equilateral triangular
cross-section: $L = 1$, $G = 1$*

n_s	Number of nodes	Number of elements	Torsional stiffness True	Torsional stiffness Computed	Error (per cent)
5	15	16		0.01522	30
9	45	64	0.02165	0.01998	7.7
13	91	144		0.02090	3.4

principle established in section 3.7.3 that the present finite element method provides upper bounds to the true stiffnesses. The principle is only applicable, however, to problems formulated with displacements or velocities as the unknowns. In this case the stress function is the unknown, and the computed stiffnesses provide lower bounds.

The program can be readily adapted to find the torsional stiffness of the cross-section shown in figure 5.11b, using the same triangular mesh (with $n_s = 13$, say) and modifying subprogram PHI1F to define zero values of ϕ_1 for the elements within the region removed. This has the effect of assigning a zero value of shear modulus to these elements. The data supplied to BCS must also be modified to ensure that zero values of the stress function are applied at the new boundary nodes. While it is somewhat wasteful to have some elements in the mesh which do not contribute to the solution, this form of mesh modification is often useful for solving problems with complicated boundary shapes. The effect of changing the section shown in figure 5.11a to that shown in figure 5.11b is to reduce the torsional stiffness by an estimated 80 per cent.

The examples described in this section demonstrate the ease with which the finite element method may be applied to problems with relatively complicated boundary shapes, which would be more difficult to solve by the finite difference method described in section 2.4.1. Note that there is no need to resort to co-ordinate systems other than cartesian. The reader is now equipped with a method and set of computer subprograms for solving any problem of the harmonic type.

6 Finite Element Analysis of Biharmonic Problems

In this chapter the formulation of a finite element analysis for two-dimensional problems of the biharmonic plane strain or plane stress types outlined in chapter 2 is described. Provision is made for the inclusion of both thermal strains and body forces, the effect of which on the governing differential equations is to modify either equation 2.72 or 2.78 to the more general form of equation 2.88. As with harmonic problems, the ϕ function in the governing differential equation is associated with externally applied forces: changes in temperature are equivalent to the application of such forces.

Although the emphasis here is on plane strain and plane stress problems, in principle the analysis and resulting computer program can be applied to other problems of the biharmonic type. Because displacements are treated as the unknowns, however, the method is unsuitable for plane strain problems involving incompressible materials. For example, it cannot be applied to the recirculating viscous flow problem outlined in section 2.2.7. Attention is concentrated on the simple CST type of finite elements. There is a close similarity of approach between the present direct equilibrium formulation and those described in chapters 1 and 3 for the analysis of structures and harmonic problems respectively.

6.1 Derivation of the Element Stiffness Matrix

Figure 3.1 again provides an illustration of a two-dimensional solution domain divided into a mesh of triangular elements. Since the present analysis is for elastic solid continua, the domain represents a solid body of which the elements are material subregions.

6.1.1 Element geometry and choice of shape function The numbering of nodes and elements, and the definitions of element dimensions and areas are exactly as described in section 3.1.1. Linear shape functions for the unknowns, in this case displacements, are also appropriate. The main difference is that here

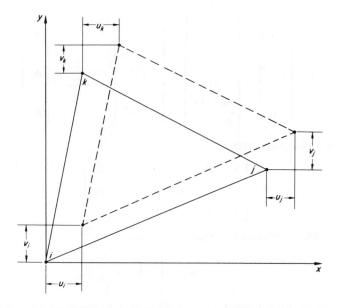

Figure 6.1   Displacement of the typical element

the displacements are in the plane of, rather than normal to, the solution domain. Figure 6.1 shows the nodal point displacements for the typical element. The displacements at points within the element are given by

$$u(x, y) = C_1 + C_2 x + C_3 y \tag{6.1}$$

$$v(x, y) = C_4 + C_5 x + C_6 y \tag{6.2}$$

where C_1 to C_6 are constant for the particular element, and x and y are local co-ordinates with the origin at node i. Now $C_1 = u_i$, $C_4 = v_i$ and the remaining parameters may be found with the aid of equations 3.11 and 3.12 as

$$\begin{bmatrix} C_2 & C_5 \\ C_3 & C_6 \end{bmatrix} = \frac{1}{2\Delta_m} \begin{bmatrix} b_i & b_j & b_k \\ a_i & a_j & a_k \end{bmatrix} \begin{bmatrix} u_i & v_i \\ u_j & v_j \\ u_k & v_k \end{bmatrix} \tag{6.3}$$

The truncation errors involved in using linear shape functions for u and v are of the same form as that shown for w in equation 3.79.

The analyses of plane strain and plane stress given in sections 2.2.5 and 2.2.6 involve only the strain components e_{xx}, e_{yy} and e_{xy}. Using the strain definitions given in equations 2.2 and 2.3 these may be expressed as

$$
e = \begin{bmatrix} e_{xx} \\ e_{yy} \\ e_{xy} \end{bmatrix} = \begin{bmatrix} \dfrac{\partial u}{\partial x} \\ \dfrac{\partial v}{\partial y} \\ \dfrac{\partial u}{\partial y} + \dfrac{\partial v}{\partial x} \end{bmatrix} = \begin{bmatrix} C_2 \\ C_6 \\ C_3 + C_5 \end{bmatrix} = \frac{1}{2\Delta_m} B \begin{bmatrix} \delta_i \\ \delta_j \\ \delta_k \end{bmatrix} \qquad (6.4)
$$

where B is a dimension matrix

$$
B = \begin{bmatrix} b_i & 0 & b_j & 0 & b_k & 0 \\ 0 & a_i & 0 & a_j & 0 & a_k \\ a_i & b_i & a_j & b_j & a_k & b_k \end{bmatrix} \qquad (6.5)
$$

and displacement terms such as δ_i are subvectors

$$
\delta_i = \begin{bmatrix} u_i \\ v_i \end{bmatrix} \qquad (6.6)
$$

The present notation follows the general form used previously: the dimension matrix B is analogous to the one defined in equation 3.12, and the symbol δ is used for generalised displacements or velocities as in equations 1.13 and 3.23.

Since the analysis is formulated with displacements as the unknowns, compatibility of strains as defined by equations 2.14 to 2.19 is automatically satisfied within each element. The displacements are also continuous across the inter-element boundaries. Linear shape functions ensure that no holes or overlaps occur between elements and a typical pair of elements is displaced and deformed as shown in figure 6.2a. Elements of the nonconforming or incompatible type discussed in section 3.7.1 tend to deform as shown in figure 6.2b.

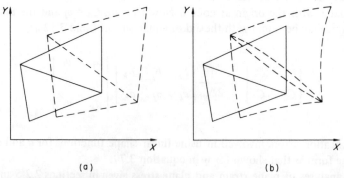

(a)   (b)

Figure 6.2   Displacement and deformation of a pair of elements (dotted lines show deformed shapes): (a) displacement continuous across inter-element boundaries; (b) displacement discontinuous

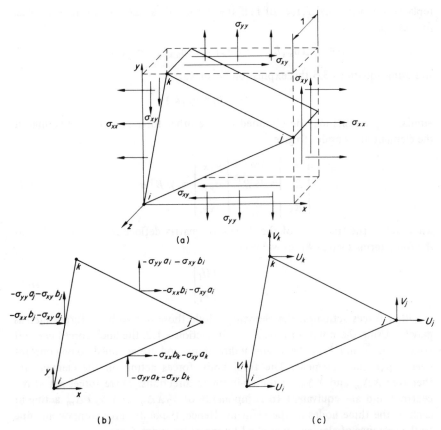

Figure 6.3   Equivalent systems of stresses and forces acting on an element: (a) stresses on
the enclosing prism; (b) forces at the mid-points of the element sides;
(c) forces at the nodes

6.1.2 Forces acting on the element   The strains and therefore the corresponding stresses σ_{xx}, σ_{yy} and σ_{xy} are constant over each element. Figure 6.3a shows these stresses acting on the rectangular prism of unit thickness enclosing the typical element shown in figure 3.2. Their effects can be expressed in terms of equivalent forces acting at the mid-points of the sides of the element as shown in figure 6.3b. For example, consider the side joining nodes i and j. The force in the x-direction at the mid-point of this side is due to a stress of $+\sigma_{xx}$ acting on an area of $(-b_k) \times 1$, together with a stress of $-\sigma_{xy}$ acting on an area of $a_k \times 1$, and is therefore equal to $-\sigma_{xx}b_k - \sigma_{xy}a_k$.

A further transformation allows the forces acting at the mid-points of the sides of the element to be replaced by an equivalent set acting at the nodes as shown in figure 6.3c. In order to maintain the same resultant force and moment about any point on a side of the element, the force at the mid-point must be replaced by two equal forces of half the magnitude at the relevant nodes. Thus, for example

$$U_i = -\tfrac{1}{2}(\sigma_{xx}b_k + \sigma_{xy}a_k) - \tfrac{1}{2}(\sigma_{xx}b_j + \sigma_{xy}a_j)$$

and using equations 3.3 this expression becomes

$$U_i = \tfrac{1}{2}(\sigma_{xx}b_i + \sigma_{xy}a_i)$$

Similar expressions may be obtained for the other force components acting on the element at its nodes to give

$$
\begin{bmatrix} R_i \\ R_j \\ R_k \end{bmatrix} = \tfrac{1}{2} B^{\mathrm{T}} \begin{bmatrix} \sigma_{xx} \\ \sigma_{yy} \\ \sigma_{xy} \end{bmatrix} = \tfrac{1}{2} B^{\mathrm{T}} \boldsymbol{\sigma}
\tag{6.7}
$$

where B^{T} is the transpose of the dimension matrix defined in equation 6.5, and the force terms such as R_i are subvectors

$$
R_i = \begin{bmatrix} U_i \\ V_i \end{bmatrix}
\tag{6.8}
$$

Other forces acting on the element include those due to body forces such as gravity. Using the notation introduced in section 2.1.2, the body forces per unit volume are $\bar{X}$ and $\bar{Y}$ in the x- and y-directions, and are assumed to be constant over a particular element. The total body forces acting on the element are therefore $\bar{X}\Delta_m$ and $\bar{Y}\Delta_m$ in the co-ordinate directions. These forces act at the centroid and are equivalent to components of $\tfrac{1}{3}\bar{X}\Delta_m$ and $\tfrac{1}{3}\bar{Y}\Delta_m$ acting at each of the three nodes of the element. Hence, the body force components due to the presence of element m applied to one of its nodes, i, are

$$
G_i^{(m)} = \frac{\Delta_m}{3} \begin{bmatrix} \bar{X} \\ \bar{Y} \end{bmatrix}_m
\tag{6.9}
$$

6.1.3 Constitutive equations   The relationships between stresses and strains may be obtained with the aid of constitutive equations 2.21 to 2.24 for an elastic solid. If the plane stress condition described in section 2.2.6 is assumed then $\sigma_{zz} = 0$ and

$$
\begin{bmatrix} e_{xx} \\ e_{yy} \\ e_{xy} \end{bmatrix} = \frac{1}{E} \begin{bmatrix} 1 & -\nu & 0 \\ -\nu & 1 & 0 \\ 0 & 0 & 2(1+\nu) \end{bmatrix} \begin{bmatrix} \sigma_{xx} \\ \sigma_{yy} \\ \sigma_{xy} \end{bmatrix} + \alpha\,\Delta T \begin{bmatrix} 1 \\ 1 \\ 0 \end{bmatrix}
\tag{6.10}
$$

The temperature change ΔT must be a function of x and y only for the problem to remain two-dimensional, and for present purposes a single average value is used for each element. Equations 6.10 may be inverted to give stresses in terms of strains in the following generalised form valid for either plane stress or plane strain

$$\sigma = D(e - e_T) \tag{6.11}$$

The vector of thermal strains, e_T, and the elastic property matrix are defined as

$$e_T = \alpha^* \, \Delta T \begin{bmatrix} 1 \\ 1 \\ 0 \end{bmatrix}, \quad D = \frac{E^*}{1-\nu^{*2}} \begin{bmatrix} 1 & \nu^* & 0 \\ \nu^* & 1 & 0 \\ 0 & 0 & \tfrac{1}{2}(1-\nu^*) \end{bmatrix} \tag{6.12}$$

where for plane stress

$$E^* = E, \quad \nu^* = \nu, \quad \alpha^* = \alpha \tag{6.13}$$

The plane strain condition described in section 2.2.5 is rather more complex if temperature changes are involved. If the plane strain assumption $e_{zz} = 0$ holds, then tensile stresses are generated in the z-direction by temperature changes alone. From equation 2.23

$$\sigma_{zz} = \nu(\sigma_{xx} + \sigma_{yy}) - E \, \alpha \, \Delta T \tag{6.14}$$

and the constitutive equations become

$$\begin{bmatrix} e_{xx} \\ e_{yy} \\ e_{xy} \end{bmatrix} = \frac{1}{E} \begin{bmatrix} (1-\nu^2) & -\nu(1+\nu) & 0 \\ -\nu(1+\nu) & (1-\nu^2) & 0 \\ 0 & 0 & 2(1+\nu) \end{bmatrix} \begin{bmatrix} \sigma_{xx} \\ \sigma_{yy} \\ \sigma_{xy} \end{bmatrix} + \alpha(1+\nu)\Delta T \begin{bmatrix} 1 \\ 1 \\ 0 \end{bmatrix} \tag{6.15}$$

Inversion of these equations yields equations 6.11 and 6.12, where for plane strain the modified material properties are

$$E^* = \frac{E}{1 - \nu^2}, \quad \nu^* = \frac{\nu}{1 - \nu}, \quad \alpha^* = (1 + \nu)\alpha \tag{6.16}$$

An alternative assumption is that thermal strains but not elastic strains are permitted in the z-direction, the result of which is to make $\alpha^* = \alpha$.

In the case of plane strain the common factor involved in the coefficients of the elastic property matrix is

$$\frac{E^*}{1 - \nu^{*2}} = \frac{E(1 - \nu)}{(1 + \nu)(1 - 2\nu)} \tag{6.17}$$

which becomes infinite when $\nu = \frac{1}{2}$. According to equation 2.27 this condition represents an incompressible material, for which the present finite element method is unsuitable.

The general constitutive equation 6.11 and the strain definitions given in equations 6.4 may be substituted into equations 6.7 to enable the forces acting on the element at the nodes due to the internal stresses to be expressed in terms of the corresponding displacements

$$\begin{bmatrix} R_i \\ R_j \\ R_k \end{bmatrix} = \frac{1}{4\Delta_m} B^T D B \begin{bmatrix} \delta_i \\ \delta_j \\ \delta_k \end{bmatrix} - \tfrac{1}{2} B^T D e_T \tag{6.18}$$

This result may be expressed in the general notation introduced in sections 1.2.2 and 3.1.3 as

$$R_m = k_m \delta_m - \theta_m \qquad (6.19)$$

where δ_m is the element displacement vector and θ_m is effectively a 'thermal force' vector for the element due to its change in temperature. The element stiffness matrix is given by

$$k_m = \frac{1}{4\Delta_m} B^T D B = \begin{bmatrix} k_{11} & k_{12} & k_{13} \\ k_{21} & k_{22} & k_{23} \\ k_{31} & k_{32} & k_{33} \end{bmatrix}_m \qquad (6.20)$$

where each of the coefficients is a 2 x 2 submatrix which can be expressed as

$$k_{rs} = \begin{bmatrix} k_{xx} & k_{xy} \\ k_{yx} & k_{yy} \end{bmatrix}_{rs} \qquad (6.21)$$

For example, k_{xy} in this submatrix can be interpreted as the force which must be applied in the x-direction to the element at the node corresponding to the rth row of R_m (that is, i, j or k according to whether r is 1, 2 or 3) to cause a unit displacement in the y-direction at the node corresponding to the sth row of δ_m.

By using submatrices in this way, the numbers of rows and columns of submatrices are kept equal to the number of nodal points. In contrast, in the alternative approach used in section 1.2.2, matrices are displayed in full with twice as many rows and columns as there are nodal points. In the harmonic analysis described in chapter 3 the distinction does not arise because there is only one unknown at each node. For present purposes both approaches have their advantages and disadvantages, mainly in terms of ease of programming and depending on which method of solving the overall linear algebraic equations is employed. The submatrix approach is chosen to maintain the closest similarity with the method and program for harmonic problems. Nevertheless, in the program described in section 6.6 the element stiffness matrix is first generated in its full 6 x 6 form.

The form of equation 6.20 ensures that k_m is symmetric provided D is symmetric. Note the similarity to equation 3.20: the matrix D is replaced by the single material parameter μ.

6.2   Assembly of the Overall Stiffness Matrix

The actual internal stresses and body forces acting on individual elements have been replaced by the equivalent forces acting at the nodes of the mesh. The conditions required for equilibrium can be expressed as

$$\sum \begin{pmatrix} \text{externally applied} \\ \text{forces at the nodes} \end{pmatrix} = \sum \begin{pmatrix} \text{forces on the elements} \\ \text{at these nodes} \end{pmatrix}$$

For example, for equilibrium of forces acting at node i

$$F_i + \Sigma G_i^{(m)} = \Sigma R_i^{(m)} \tag{6.22}$$

where the subvector F_i represents the forces applied externally at the node

$$F_i = \begin{bmatrix} F_x \\ F_y \end{bmatrix}_i \tag{6.23}$$

Such forces usually take the form of surface tractions, as described in section 6.5.1, and exclude body forces. The body forces $G_i^{(m)}$ and the internal forces $R_i^{(m)}$ applied to the elements are defined according to equations 6.9 and 6.18. The summations indicated in equation 6.22 are performed for elements which have the point i as a node.

The set of equations for equilibrium of all the nodes can be expressed in the general form

$$K\delta = F^* \tag{6.24}$$

where

$$K\delta = \Sigma k_m \delta_m \tag{6.25}$$

and

$$F^* = F + G + \theta \tag{6.26}$$

As usual, K and δ are the overall stiffness matrix and displacement vectors respectively. The vector F^* represents the externally applied forces modified for both body and thermal forces contained in the overall vectors G and θ. Equation 6.23 displays a typical subvector of F, while those of G and θ are given by

$$G_i = \Sigma G_i^{(m)}, \quad \theta_i = \Sigma \theta_i^{(m)} \tag{6.27}$$

The subvector $\theta_i^{(m)}$ represents the thermal forces at the point i due to the presence of element m, and is obtained from the element thermal force vector defined by equations 6.18 and 6.19.

The process of assembling the overall stiffness matrix is as described in section 3.2 for harmonic problems, except that instead of single coefficients, 2 x 2 submatrices are assembled according to equation 3.26. Expressed in terms of individual stiffness coefficients this becomes

$$\begin{bmatrix} K_{xx} & K_{xy} \\ K_{yx} & K_{yy} \end{bmatrix}_{pq} = \Sigma \begin{bmatrix} k_{xx} & k_{xy} \\ k_{yx} & k_{yy} \end{bmatrix}_{rs} \tag{6.28}$$

As the individual element stiffness matrices are symmetric then so is the overall stiffness matrix.

6.3   Variational Formulation

As for harmonic problems, a variational formulation of the finite element analysis for biharmonic problems provides a more general approach than the above direct equilibrium method. The variational method described in section 3.4.1, in which the variational principle is derived from the governing differential equation, could be applied to plane strain and plane stress problems. Since for present purposes displacements rather than a stress function are treated as the unknowns, the relevant differential equations would be equilibrium equations 2.7 and 2.8 (with stresses expressed in terms of derivatives of displacements) rather than equations 2.72 or 2.78.

Alternatively, the required functional χ can be taken as the total potential energy of the solution domain, which is stationary when equilibrium is achieved. That is

$$\chi = \begin{pmatrix} \text{strain energy} \\ \text{stored} \end{pmatrix} + \begin{pmatrix} \text{potential energy of the} \\ \text{external forces} \end{pmatrix} \qquad (6.29)$$

For small changes $d\delta$ in the displacements, the corresponding change in χ for the present two-dimensional problems is

$$d\chi = \iint (\sigma_{xx} de_{xx} + \sigma_{yy} de_{yy} + \sigma_{xy} de_{xy})\, dx\, dy$$
$$- \iint (\bar{X} du + \bar{Y} dv)\, dx\, dy - F^{\mathrm{T}} d\delta \qquad (6.30)$$

where the integrations are carried out over the entire solution domain. The external forces F are assumed to be applied only at the nodes of the finite element mesh: the application of distributed forces is discussed in section 6.5.1. Assuming the elements employed are of the conforming type, such that equation 3.50 is applicable, equation 6.30 can be expressed in terms of element matrices as

$$d\chi = \Sigma \Delta_m \sigma^{\mathrm{T}}\, de - \Sigma G_m^{\mathrm{T}}\, d\delta_m - F^{\mathrm{T}} d\delta \qquad (6.31)$$

where the summations are carried out over all the elements in the mesh.

The body force term deserves some explanation. The mean displacements within the typical element are

$$\bar{u} = \tfrac{1}{3}(u_i + u_j + u_k), \quad \bar{v} = \tfrac{1}{3}(v_i + v_j + v_k)$$

and if the body forces $\bar{X}$ and $\bar{Y}$ are constant over the element

$$\iint (\bar{X}\, du + \bar{Y}\, dv)\, dx\, dy = \Sigma \Delta_m (\bar{X}\, d\bar{u} + \bar{Y}\, d\bar{u})$$
$$= \Sigma \tfrac{1}{3} \Delta_m (\bar{X}\, du_i + \bar{Y}\, dv_i + \bar{X}\, du_j + \bar{Y}\, dv_j + \bar{X}\, du_k + \bar{Y}\, dv_k)$$
$$= \Sigma G_m^{\mathrm{T}}\, d\delta_m$$

With elements of a more sophisticated type than CST, more complicated integrations are required for the evaluation of both the work done by the external forces and the strain energy stored.

Using equation 6.4 for e, equation 6.31 becomes

$$d\chi = \Sigma \tfrac{1}{2}\sigma^T B \, d\delta_m - \Sigma G_m^T \, d\delta_m - F^T d\delta$$

Since the required solution is obtained when the value of χ is stationary with respect to the nodal point displacements, $d\chi = 0$ and

$$\Sigma \tfrac{1}{2}\sigma^T B = \Sigma G_m^T + F^T$$

Each of the terms in this equation may be transposed to give

$$\Sigma \tfrac{1}{2} B^T \sigma = F + G$$

and σ may be defined in terms of displacements with the aid of equations 6.11 and 6.4 to give

$$\Sigma \frac{1}{4\Delta_m} B^T D B \delta_m = F + G + \Sigma \tfrac{1}{2} B^T D e_T \tag{6.32}$$

which is identical with equation 6.24.

In deriving equation 6.31 from equation 6.30 it is assumed that the interfaces between the elements make no contribution to χ. For CST elements the criterion established in section 3.7.1 for this to be true is that χ should involve derivatives of the unknowns of no higher than the first order. The products of stresses and strains are the only terms in χ to involve derivatives of the displacements, and these are of the first order. Hence the present finite element method is convergent in the sense defined in section 3.7. Following the argument outlined in section 3.7.3, the computed stiffnesses provide upper bounds to the true stiffnesses.

6.4 Solution of the Linear Equations

Although boundary conditions must be applied before the overall linear algebraic equations 6.24 are solved, for present purposes it is convenient to consider the solution process first. The methods of applying displacement boundary conditions described in section 6.5.2 require prior knowledge of the method of solution.

The general discussion of methods of solution presented in section 3.6 for harmonic problems is equally applicable to biharmonic problems. The main difference is that single matrix and vector coefficients are replaced by submatrices and subvectors: simple arithmetic operations are replaced by the equivalent matrix operations. Although the storage requirements for overall vectors and matrices are multiplied by two and four respectively, the relative storage requirements and computational efficiencies of the direct and iterative methods of solution remain unchanged.

For the reasons discussed in section 3.6.3, attention is concentrated on the Gauss—Seidel method. Following equation 3.71, the changes in the unknown displacements between successive cycles of iteration are computed from equations 6.24 as

$$\Delta \boldsymbol{\delta}_i = K_{ii}^{-1} \left(F_i^* - \sum_{j=1}^{n} K_{ij} \boldsymbol{\delta}_j \right) \tag{6.33}$$

where the $\boldsymbol{\delta}_i$ and F_i^* are now subvectors and the K_{ij} are submatrices. If K is stored in the rectangular form illustrated by equations 3.69 and 3.70 then as in equation 3.72

$$\Delta \delta_i = \tilde{K}_{i1}^{-1} \left(F_i^* - \sum_{l=1}^{L_i} \tilde{K}_{il} \delta_j \right) \tag{6.34}$$

where $j = M_{il}$. The rectangularised overall stiffness matrix $\tilde{K}$ is also composed of submatrices, while the corresponding single coefficients of M store the original column numbers in K. The coefficients of vector L store the numbers of nonzero submatrices in the corresponding rows of either K or $\tilde{K}$.

Now $\tilde{K}_{i1}^{-1}$ is the inverse of the self-stiffness submatrix for node i and is referred to as the self-flexibility submatrix. Its coefficients may be obtained from the self-stiffness coefficients as follows

$$\tilde{K}_{i1}^{-1} = \begin{bmatrix} \tilde{K}_{xx} & \tilde{K}_{xy} \\ \tilde{K}_{yx} & \tilde{K}_{yy} \end{bmatrix}_{i1}^{-1} = f_i = \begin{bmatrix} f_{xx} & f_{xy} \\ f_{yx} & f_{yy} \end{bmatrix}_i \tag{6.35}$$

where

$$f_{xx} = \frac{\tilde{K}_{yy}}{|\tilde{K}_{i1}|}, \ f_{xy} = \frac{-\tilde{K}_{xy}}{|\tilde{K}_{i1}|}, \ f_{yx} = \frac{-\tilde{K}_{yx}}{|\tilde{K}_{i1}|}, \ f_{yy} = \frac{\tilde{K}_{xx}}{|\tilde{K}_{i1}|} \tag{6.36}$$

and the determinant of the self-stiffness submatrix is

$$|\tilde{K}_{i1}| = \tilde{K}_{xx}\tilde{K}_{yy} - \tilde{K}_{xy}\tilde{K}_{yx} \tag{6.37}$$

In section 3.6.2 a considerable amount of attention is devoted to the convergence of the Gauss—Seidel method applied to harmonic problems. The sufficient condition for convergence is that of diagonal dominance of the overall stiffness matrix. For harmonic problems this condition is achieved if there are no obtuse-angled elements so that every element stiffness matrix is diagonally dominant. Turning to biharmonic problems, the 6 x 6 element stiffness matrix defined by equation 6.20 is never diagonally dominant and difficulties may be experienced in achieving convergence. Without attempting to present a detailed analysis of the stiffness matrix, it can be stated that obtuse-angled elements should again be avoided and that the best conditions for convergence are obtained when the elements are as nearly equilateral as possible. Long thin elements are to be avoided as strenuously as angles greater than a right angle. Given a reasonable mesh, however, the convergence of the Gauss—Seidel method applied to biharmonic problems is generally satisfactory.

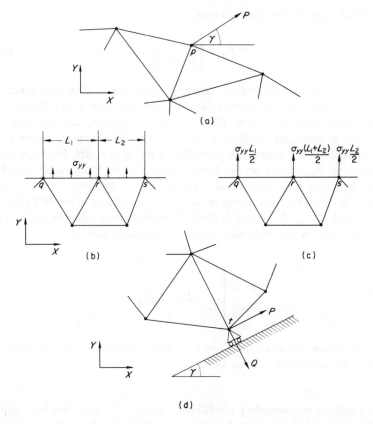

Figure 6.4   Types of boundary conditions: (a) point force; (b) distributed force (surface stress); (c) equivalent nodal point forces; (d) node restrained to move freely in a particular direction

6.5   Boundary Conditions

Boundary conditions encountered in plane strain and plane stress problems include prescribed surface tractions and restraints applied to some of the displacements. For example, the points on a particular boundary may be given prescribed displacements, or may be allowed to move freely in a prescribed direction, usually parallel to the boundary. Some practical examples of boundary conditions are provided by the case studies presented in chapter 7.

6.5.1 Force boundary conditions   Surface tractions may be specified in terms of prescribed forces acting at particular points. For example, a force P acting at an angle γ to the X global co-ordinate axis might be applied to node p as shown in figure 6.4a. Such a condition can be accommodated by defining the subvector of externally applied forces at the node as

$$F_p = \begin{bmatrix} P \cos \gamma \\ P \sin \gamma \end{bmatrix} \tag{6.38}$$

Clearly, if point force boundary conditions are to be applied, it is desirable to employ a mesh which has nodes at the points of application of the forces.

Distributed external forces (that is, surface stresses) are also frequently prescribed as boundary conditions. At least for CST elements, however, they may be replaced by equivalent point forces acting at nodes. For example, the tensile stress σ_{yy} acting on the horizontal surface between nodes q and s as shown in figure 6.4b may be replaced by the nodal point forces shown in figure 6.4c. The total force acting on, say, the element side of length L_1 between nodes q and r is $\sigma_{yy}L_1$, which is divided equally between the two nodes. Hence for the point r, say, the prescribed externally applied forces are

$$F_r = \begin{bmatrix} 0 \\ \sigma_{yy} \dfrac{(L_1 + L_2)}{2} \end{bmatrix} \tag{6.39}$$

Nonuniform stress distributions can also be accommodated by taking average values of stress over the element sides.

6.5.2 Displacement boundary conditions　　Section 3.5 describes how displacement (velocity) boundary conditions may be applied in harmonic problems by modifying the overall stiffness matrix. Equivalent approaches can be used in the present context of biharmonic problems. Since equation 6.34 is used to compute changes in the displacements between successive cycles of Gauss–Seidel iteration, displacements prescribed for node i are retained throughout the solution process if $\Delta \delta_i = 0$. This can be achieved by the following modification to the self-flexibility submatrix

$$f_i^* = \begin{bmatrix} 0 & 0 \\ 0 & 0 \end{bmatrix} \tag{6.40}$$

The other type of displacement boundary condition to be accommodated occurs when points on a particular boundary are restrained to move freely in a prescribed direction. Axes of symmetry provide the most common examples of such conditions. Suppose the node t shown in figure 6.4d is restrained to move at an angle γ to the X-axis. In general there may be a prescribed force P acting on the node in this direction. Another force, Q say, is required in the direction normal to the motion in order to maintain the restraint. Therefore, the externally applied forces at the node are

$$F_t = \begin{bmatrix} Q \sin \gamma + P \cos \gamma \\ -Q \cos \gamma + P \sin \gamma \end{bmatrix} \tag{6.41}$$

Since the value of Q is not known in advance, these components cannot be prescribed as force boundary conditions. Now from equations 6.34, 6.35 and 6.26

$$\Delta \delta_t = \begin{bmatrix} \Delta u_t \\ \Delta v_t \end{bmatrix} = \begin{bmatrix} f_{xx} & f_{xy} \\ f_{yx} & f_{yy} \end{bmatrix}_t \begin{bmatrix} Q \sin \gamma + S_x \\ -Q \cos \gamma + S_y \end{bmatrix} \tag{6.42}$$

where

$$\begin{bmatrix} S_x \\ S_y \end{bmatrix} = \begin{bmatrix} P \cos \gamma \\ P \sin \gamma \end{bmatrix} + G_t + \theta_t - \sum_{l=1}^{L_t} \tilde{K}_{tl} \, \delta_j \tag{6.43}$$

and

$$j = M_{tl}$$

The self-flexibility coefficients can be modified to eliminate the unknown Q as follows

$$\begin{bmatrix} \Delta u_t \\ \Delta v_t \end{bmatrix} = \begin{bmatrix} f_{xx}^* & f_{xy}^* \\ f_{yx}^* & f_{yy}^* \end{bmatrix}_t \begin{bmatrix} S_x \\ S_y \end{bmatrix} \tag{6.44}$$

Since $\Delta v_t = \Delta u_t \tan \gamma$, the modified flexibility coefficients can be obtained after some algebraic manipulation as

$$f_{xx}^* = \frac{f_{xx}f_{yy} - f_{xy}f_{yx}}{f_{xx} \tan^2 \gamma - (f_{xy} + f_{yx}) \tan \gamma + f_{yy}} \tag{6.45}$$

$$f_{xy}^* = f_{yx}^* = f_{xx}^* \tan \gamma \tag{6.46}$$

$$f_{yy}^* = f_{xx}^* \tan^2 \gamma \tag{6.47}$$

For the special case of $\gamma = 90°$, the magnitude of $\tan \gamma$ is infinite and the expressions for the modified flexibilities become

$$f_{xx}^* = f_{xy}^* = f_{yx}^* = 0 \tag{6.48}$$

$$f_{yy}^* = f_{yy} - f_{xy}f_{yx}/f_{xx} \tag{6.49}$$

A mistake commonly made in using the present finite element method is that of failing to specify sufficient boundary conditions to preclude rigid body motion. Usually it is necessary to prescribe zero displacements for at least one node in the mesh and restrained direction of motion for at least one other node. If, for example, only force boundary conditions are specified, then even though the external forces on the body are in equilibrium the displacements are indeterminate. Attempts to solve such a problem result in either failure of the

Gauss—Seidel method to converge or the detection of a singular overall stiffness matrix by a direct elimination procedure (see appendix A).

6.6   A Computer Program for Problems of the Biharmonic Plane Strain or Plane Stress Type

Having presented a finite element analysis for two-dimensional problems of the biharmonic type, it now remains to express the method in the form of a computer program. Some practical applications are described in chapter 7. The program is very similar in terms of general layout and variable names to the one described in chapter 3 for harmonic problems. Equivalent subprograms have identical names and at least some of them are interchangeable between the two types of problems.

Figure 6.5 shows the main program and figures 6.6, 6.7 and 6.8 show subprograms named respectively MATLS, TEMPS and BODYF for defining material properties, temperature changes and body forces. Figure 6.9 shows a subprogram named MSHOUT for writing out the mesh data and figure 6.10 a subprogram named BCS for applying the relevant boundary conditions. Figure 6.11 shows a subprogram named SOLVE2 for solving the overall linear equations by the Gauss—Seidel method, and figure 6.12 a subprogram named OUTPUT for writing out the results of the computation. Other subprograms called by the main program are MESH and MODIFY which provide the mesh data as described in chapter 4.

6.6.1 The main program   As far as possible the program variable names used in figure 6.5 are the same as those used in the other programs presented in this book. The geometric data for the mesh are stored in arrays which are located in the COMMON block of storage named CMESH, and are as described in section 3.8.1.

The arrays of variables used in the solution of the overall set of linear equations are located in the COMMON block of storage named CEQNS. The overall stiffness submatrix coefficients K_{xx}, K_{xy}, K_{yx} and K_{yy} are stored in the arrays OKXX, OKXY, OKYX and OKYY. The form of storage of these stiffnesses is rectangular, arrays NPA and NAP being used as described in section 3.8.1 to store the numbers assigned to adjacent nodal points and the total number of such points respectively. The arrays U and V are used to store the nodal point displacement components which form the vector δ in equations 6.24. Similarly, FX and FY store the components of the forces applied externally to the nodes, while FXMOD and FYMOD store their values modified for both body and thermal forces. The arrays SFXX, SFXY, SFYX and SFYY are used to store the coefficients f_{xx}, f_{xy}, f_{yx} and f_{yy} of the self-flexibility submatrices for the nodal points, either as defined in equations 6.36 or subsequently modified for some of the boundary points as in equations 6.40, 6.45 to 6.47, or 6.48 and 6.49.

```
C  PROGRAM FOR FINITE ELEMENT ANALYSIS OF TWO-DIMENSIONAL PROBLEMS OF
C  THE BIHARMONIC PLANE STRAIN OR PLANE STRESS TYPE, USING CONSTANT
C  STRAIN TRIANGULAR ELEMENTS.
C
       DIMENSION  TITLE(6),B(3,6),D(3,3),BTD(6,3),ESTIFF(6,6),IJK(3),
      1            ET(3),THETAM(6)
       REAL  NU
       COMMON /CMESH/ NEL,NNP,X(121),Y(121),AI(200),AJ(200),AK(200),
      1     BI(200),BJ(200),BK(200),AREA(200),NPI(200),NPJ(200),NPK(200),
      2     NBP,NPB(40),MOUT
      3        /CEQNS/ OKXX(121,9),OKXY(121,9),OKYX(121,9),OKYY(121,9),
      4     U(121),V(121),FX(121),FY(121),FXMOD(121),FYMOD(121),SFXX(121),
      5     SFXY(121),SFYX(121),SFYY(121),NPA(121,9),NAP(121)
      6        /CMATL/ NMAT,E(5),NU(5),ALPHA(5),RHO(5),MATM(200)
      7        /CLOAD/ DELTAT(200),XBAR(200),YBAR(200)
       DATA  BLANK,STRESS,STRAIN / 10H          ,6HSTRESS,6HSTRAIN /
C
C  INPUT THE PROBLEM TITLE AND TYPE - STOP IF BLANK CARD ENCOUNTERED.
  1    READ(5,51) TITLE,CASE
 51    FORMAT(6A10,A6)
       IF(TITLE(1).EQ.BLANK) STOP
       WRITE(6,61) CASE,TITLE
 61    FORMAT(39H0CST FINITE ELEMENT SOLUTION FOR PLANE ,A6,8H PROBLEM
      1     // 6A10)
C
C  INPUT OR GENERATE THE MESH DATA, MATERIAL PROPERTIES, TEMPERATURE
C  CHANGES AND BODY FORCES.
       CALL   MESH
       CALL   MODIFY
       CALL   MATLS(NEL)
       CALL   TEMPS(NEL)
       CALL   BODYF(NEL)
C
C  COMPUTE THE ELEMENT GEOMETRIES.
       DO 2 M=1,NEL
       I=NPI(M)
       J=NPJ(M)
       K=NPK(M)
       AI(M)=-X(J)+X(K)
       AJ(M)=-X(K)+X(I)
       AK(M)=-X(I)+X(J)
       BI(M)=Y(J)-Y(K)
       BJ(M)=Y(K)-Y(I)
       BK(M)=Y(I)-Y(J)
       AREA(M)=0.5*(AK(M)*BJ(M)-AJ(M)*BK(M))
       IF(AREA(M).GT.0.) GO TO 2
       WRITE(6,62) M
 62    FORMAT(15HOELEMENT NUMBER,I5,25H HAS NEGATIVE AREA - STOP)
       STOP
  2    CONTINUE
C
C  OUTPUT THE MESH DATA.
       CALL   MSHOUT
C
C  SET INITIAL VALUES OF STIFFNESSES, EXTERNAL FORCES AND UNKNOWNS.
       DO 4 IROW=1,NNP
       DO 3 IC=1,9
       OKXX(IROW,IC)=0.
       OKXY(IROW,IC)=0.
       OKYX(IROW,IC)=0.
       OKYY(IROW,IC)=0.
  3    NPA(IROW,IC)=0
       NPA(IROW,1)=IROW
       FXMOD(IROW)=0.
       FYMOD(IROW)=0.
       U(IROW)=0.
  4    V(IROW)=0.
```

Figure 6.5   Main program for finite element analysis of biharmonic problems of the plane
strain or plane stress types

```
C
C   MODIFY MATERIAL PROPERTIES IF CASE IS ONE OF PLANE STRAIN.
        IF(CASE.EQ.STRESS) GO TO 6
        DO 5 MAT=1,NMAT
        E(MAT)=E(MAT)/(1.-NU(MAT)**2)
        ALPHA(MAT)=ALPHA(MAT)*(1.+NU(MAT))
    5   NU(MAT)=NU(MAT)/(1.-NU(MAT))
C
C   SET UP THE OVERALL ASSEMBLY LOOP.
    6   DO 19 M=1,NEL
C
C   STORE THE ELEMENT NODE NUMBERS IN ORDER IN ARRAY IJK.
        IJK(1)=NPI(M)
        IJK(2)=NPJ(M)
        IJK(3)=NPK(M)
C
C   COMPUTE THE BODY FORCE COMPONENTS ON EACH NODE OF THE ELEMENT.
        GXM=XBAR(M)*AREA(M)/3.
        GYM=YBAR(M)*AREA(M)/3.
C
C   FORM THE ELEMENT DIMENSION MATRIX.
        DO 7 IRE=1,2
        DO 7 ICE=1,6
    7   B(IRE,ICE)=0.
        B(1,1)=BI(M)
        B(1,3)=BJ(M)
        B(1,5)=BK(M)
        B(2,2)=AI(M)
        B(2,4)=AJ(M)
        B(2,6)=AK(M)
        DO 8 ICE=1,6
        IF(MOD(ICE,2).EQ.0) B(3,ICE)=B(1,ICE-1)
    8   IF(MOD(ICE,2).EQ.1) B(3,ICE)=B(2,ICE+1)
C
C   FORM THE ELASTIC PROPERTY MATRIX.
        DO 9 IRE=1,3
        DO 9 ICE=1,3
    9   D(IRE,ICE)=0.
        MAT=MATM(M)
        FACT=E(MAT)/(1.-NU(MAT)**2)
        D(1,1)=FACT
        D(2,2)=FACT
        D(1,2)=FACT*NU(MAT)
        D(2,1)=D(1,2)
        D(3,3)=FACT*0.5*(1.-NU(MAT))
C
C   MULTIPLY THE TRANSPOSE OF MATRIX B BY MATRIX D.
        DO 10 IRE=1,6
        DO 10 ICE=1,3
        BTD(IRE,ICE)=0.
        DO 10 ISUM=1,3
   10   BTD(IRE,ICE)=BTD(IRE,ICE)+B(ISUM,IRE)*D(ISUM,ICE)
C
C   FORM THE THERMAL STRAIN AND THERMAL FORCE VECTORS.
        ET(1)=ALPHA(MAT)*DELTAT(M)
        ET(2)=ET(1)
        ET(3)=0.
        DO 12 IRE=1,6
        SUM=0.
        DO 11 ISUM=1,3
   11   SUM=SUM+BTD(IRE,ISUM)*ET(ISUM)
   12   THETAM(IRE)=0.5*SUM
C
C   FORM THE ELEMENT STIFFNESS MATRIX.
        DO 14 IRE=1,6
        DO 14 ICE=1,6
        SUM=0.
        DO 13 ISUM=1,3
   13   SUM=SUM+BTD(IRE,ISUM)*B(ISUM,ICE)
   14   ESTIFF(IRE,ICE)=0.25*SUM/AREA(M)
```

Figure 6.5   Continued

```
C
C  ADD ELEMENT STIFFNESS TO OVERALL STIFFNESS.
        DO 18 IRE=1,3
        DO 18 ICE=1,3
        IROW=IJK(IRE)
        ICOL=IJK(ICE)
C
C  STORE STIFFNESS COEFFICIENTS IN RECTANGULAR FORM OF OVERALL MATRICES.
        DO 15 IC=1,9
        IF(NPA(IROW,IC).EQ.ICOL) GO TO 17
        IF(NPA(IROW,IC).EQ.0) GO TO 16
  15    CONTINUE
        WRITE(6,63) IROW
  63    FORMAT(5H0NODE,I5,38H HAS MORE THAN 8 ADJACENT NODES - STOP)
        STOP
  16    NPA(IROW,IC)=ICOL
        NAP(IROW)=IC
  17    OKXX(IROW,IC)=OKXX(IROW,IC)+ESTIFF(2*IRE-1,2*ICE-1)
        OKXY(IROW,IC)=OKXY(IROW,IC)+ESTIFF(2*IRE-1,2*ICE)
        OKYX(IROW,IC)=OKYX(IROW,IC)+ESTIFF(2*IRE,2*ICE-1)
  18    OKYY(IROW,IC)=OKYY(IROW,IC)+ESTIFF(2*IRE,2*ICE)
C
C  ASSEMBLE THE EXTERNAL FORCES ON THE NODES.
        DO 19 IRE=1,3
        IROW=IJK(IRE)
        FXMOD(IROW)=FXMOD(IROW)+GXM+THETAM(2*IRE-1)
  19    FYMOD(IROW)=FYMOD(IROW)+GYM+THETAM(2*IRE)
C
C  COMPUTE THE SELF-FLEXIBILITY SUBMATRICES.
        DO 20 I=1,NNP
        DENOM=OKXX(I,1)*OKYY(I,1)-OKXY(I,1)*OKYX(I,1)
        SFXX(I)=OKYY(I,1)/DENOM
        SFXY(I)=-OKXY(I,1)/DENOM
        SFYX(I)=-OKYX(I,1)/DENOM
  20    SFYY(I)=OKXX(I,1)/DENOM
C
C  APPLY THE BOUNDARY CONDITIONS.
        CALL  BCS
C
C  SOLVE THE LINEAR EQUATIONS.
        CALL  SOLVE2(NNP)
C
C  OUTPUT THE REQUIRED RESULTS.
        CALL  OUTPUT
        GO TO 1
        END
```

Figure 6.5   Continued

Material, temperature rise and body force data are stored in arrays located in the COMMON blocks of storage named CMATL and CLOAD, and are discussed in sections 6.6.2, 6.6.3 and 6.6.4. The other subscripted variables used in the main program are TITLE which stores a title for the problem, B, D and BTD which store the coefficients of the matrices B, D and the product $B^T D$, ESTIFF which stores the coefficients of the element stiffness matrix, IJK which stores the node numbers for a particular element, and ET and THETAM which store the coefficients of the thermal strain and force vectors e_T and θ_m. Other variables used in the program include I, J, K, M, IROW, ICOL, IRE, ICE and IC which are used for the same purposes as in the program for harmonic problems. Note that matrix row and column numbers now refer to rows and columns of submatrices.

The first action of the main program is to read from the first data card the problem title and the type of case: the plane stress or plane strain conditions are selected by supplying either the alphanumeric data STRESS or STRAIN to be read into the variable CASE. If at least the first ten columns of the card are blank then execution is terminated. Otherwise a heading is written out, followed by the problem title. Then subprograms MESH and MODIFY are called to provide the mesh geometry data, followed by MATLS, TEMPS and BODYF to define the material properties, temperature changes and body forces. The element dimensions are computed as in the harmonic program and subprogram MSHOUT is called to write out the mesh data.

The coefficients stored in the overall stiffness and external force arrays are set to zero in preparation for the overall assembly process. The first columns of the overall stiffness arrays are made to contain the self-stiffness coefficients of the nodal points by setting the node counters stored in the first column of NPA equal to the corresponding row numbers. The displacements stored in U and V are also set to zero to serve as the initial values for the Gauss–Seidel solution process: prescribed displacement boundary conditions are applied as described in section 6.6.6. If the current case is one of plane strain, the material properties are modified according to equations 6.16.

Within the program loop for the overall assembly process the external forces and stiffness matrix are computed for each element in turn. The node numbers for the particular element are first stored in array IJK and the body force components applicable to each node are computed according to equation 6.9 and stored in GXM and GYM. The element geometry matrix is then formed according to equation 6.5, the elastic property matrix according to equations 6.12, and the product B^TD is computed and stored in array BTD. This enables the thermal force vector for the element to be computed with the aid of equations 6.18 and 6.19, having first used equations 6.12 to define the thermal strain vector. Finally, the element stiffness matrix is found from the product of B^TD and B according to equation 6.20.

The assembly of the individual element stiffness coefficients into the rectangular form of the overall stiffness matrix is carried out as described for the harmonic program, the only difference being that in place of single stiffnesses there are now four submatrix coefficients which are assembled as indicated by equation 6.28. The only place where array subscripts do not refer to numbers of rows and columns of submatrices is in the element stiffness matrices, which are more conveniently computed and stored as 6 x 6 matrices of single coefficients. The necessary conversion is readily accomplished in the final stiffness assembly statements, such as the statement numbered 18. The last action within the overall assembly loop is to add the body and thermal force components due to the particular element to the modified external forces at each of its three nodes. The unmodified external forces are added later in subprogram BCS.

In preparation for the Gauss–Seidel solution process and the application of displacement boundary conditions, the self-flexibility submatrix coefficients are computed according to equation 6.36 for each of the nodal points. Sub-programs BCS, SOLVE2 and OUTPUT are then called in turn to apply the boundary conditions, to solve the overall equations and to write out the results. Finally, control of execution in the main program returns to the first input statement for a new problem.

6.6.2 The subprogram for defining material properties The subprogram MATLS shown in figure 6.6 serves to define the properties of the materials involved in a particular problem. In the case of plane strain these are subsequently modified according to equations 6.16 in the main program. Young's moduli, Poisson's ratios, coefficients of thermal expansion and densities are stored in arrays E, NU, ALPHA and RHO, which are located in the COMMON block of storage named CMATL. With the present version of the subprogram, the solution domain may involve up to five different materials, the actual number of materials being stored in NMAT. Each element is composed of one of these materials and the element material numbers are stored in the array MATM.

The number of materials is first read in, followed by the sets of property data which are then written out. The present version of the subprogram is intended for a homogeneous solution domain in that all the element material numbers are set to one. For problems involving more than one material either some means of generating element material numbers can be found, or array MATM can be read in as data.

```
      SUBROUTINE  MATLS(NEL)
C
C  SUBPROGRAM FOR DEFINING THE MATERIAL PROPERTIES OF THE ELEMENTS.
C
      REAL  NU
      COMMON /CMATL/ NMAT,E(5),NU(5),ALPHA(5),RHO(5),MATM(200)
C
C  INPUT THE MATERIAL PROPERTIES - MAXIMUM 5 DIFFERENT MATERIALS.
      READ(5,51) NMAT
 51   FORMAT(I5)
      IF(NMAT.LE.5) GO TO 1
      WRITE(6,61) NMAT
 61   FORMAT(28HOTOO MANY MATERIALS - NMAT =,I5)
      STOP
 1    READ(5,52) (MAT,E(MAT),NU(MAT),ALPHA(MAT),RHO(MAT),N=1,NMAT)
 52   FORMAT(I5,4E15.5)
      WRITE(6,62) (MAT,E(MAT),NU(MAT),ALPHA(MAT),RHO(MAT),MAT=1,NMAT)
 62   FORMAT(20HOMATERIAL PROPERTIES //
     1 50H    MATL    E            NU        ALPHA       RHO
     2     (1X,I5,E12.4,F8.3,2E12.4))
C
C  DEFINE THE MATERIAL FOR EACH ELEMENT.
C  THIS VERSION ASSUMES ALL ELEMENTS ARE OF FIRST MATERIAL.
      DO 2 M=1,NEL
 2    MATM(M)=1
      RETURN
      END
```

Figure 6.6   Subprogram for defining material properties

```
       SUBROUTINE  TEMPS(NEL)
C
C   SUBPROGRAM FOR DEFINING MEAN TEMPERATURE CHANGES FOR THE ELEMENTS.
C   THIS VERSION READS AND ASSIGNS A UNIFORM CHANGE.
C
       COMMON /CLOAD/ DELTAT(200),XBAR(200),YBAR(200)
       READ(5,51) TEMP
    51 FORMAT(F10.0)
       DO 1 M=1,NEL
     1 DELTAT(M)=TEMP
       RETURN
       END
```

Figure 6.7   Subprogram for defining temperature changes

6.6.3 The subprogram for defining temperature changes The subprogram TEMPS shown in figure 6.7 serves to define the mean temperature changes for the elements, which are stored in the array DELTAT located in the COMMON block of storage named CLOAD. In the present version a uniform change is read into the variable TEMP and its value assigned to each of the elements. For problems involving nonuniform temperature profiles the array DELTAT can be read in or, for example, the element temperature changes can be computed with the aid of the finite element analysis described in chapter 3 applied to thermal conduction within the solution domain. If the temperature change varies within a particular element, for present purposes the value at the centroid can be used as the mean value.

6.6.4 The subprogram for defining body forces The subprogram BODYF shown in figure 6.8 serves to define the body force components $\bar{X}$ and $\bar{Y}$ for the elements, which are stored in the arrays XBAR and YBAR located in the COMMON block of storage named CLOAD. These components are the values per unit volume, that is, per unit area of the solution domain, and if necessary are averaged for each element. In the present version the only body force is assumed to be the weight of the material acting in the negative y-direction. Other types of body force commonly encountered include those due to centrifugal effects.

```
       SUBROUTINE  BODYF(NEL)
C
C   SUBPROGRAM FOR DEFINING THE BODY FORCE COMPONENTS (PER UNIT VOLUME)
C   FOR THE ELEMENTS.
C   THIS VERSION ASSUMES GRAVITY ACTS IN THE NEGATIVE Y-DIRECTION.
C
       REAL  NU
       COMMON /CMATL/ NMAT,E(5),NU(5),ALPHA(5),RHO(5),MATM(200)
      1        /CLOAD/ DELTAT(200),XBAR(200),YBAR(200)
       DO 1 M=1,NEL
       XBAR(M)=0.
       MAT=MATM(M)
     1 YBAR(M)=-RHO(MAT)
       RETURN
       END
```

Figure 6.8   Subprogram for defining body forces

6.6.5 The mesh data output subprogram   The subprogram MSHOUT shown in figure 6.9 serves to write out the geometric, temperature change and body force data for the mesh. It is very similar to the identically named subprogram described in section 3.8.2.

```
      SUBROUTINE  MSHOUT
C
C   SUBPROGRAM TO WRITE OUT THE MESH DATA.
C
      REAL  NU
      COMMON /CMESH/ NEL,NNP,X(121),Y(121),AI(200),AJ(200),AK(200),
     1       BI(200),BJ(200),BK(200),AREA(200),NPI(200),NPJ(200),NPK(200),
     2       NBP,NPB(40),MOUT
     3       /CMATL/ NMAT,E(5),NU(5),ALPHA(5),RHO(5),MATM(200)
     4       /CLOAD/ DELTAT(200),XBAR(200),YBAR(200)
      IF(MOUT.EQ.0) RETURN
C
C   OUTPUT THE NUMBER OF ELEMENTS AND NODES, AND THE NODE CO-ORDINATES.
      WRITE(6,61) NEL,NNP,(I,X(I),Y(I),I=1,NNP)
   61 FORMAT(28HOGEOMETRIC DATA FOR THE MESH //
     1       10X,21H NUMBER OF ELEMENTS =,I4 //
     2       10X,25H NUMBER OF NODAL POINTS =,I4 //
     3       25H NODAL POINT CO-ORDINATES //
     4       72H      I      X          Y         I      X      Y      I
     5   X          Y      / (3(1X,I5,2F9.4)))
C
C   OUTPUT THE ELEMENT NODE AND MATERIAL NUMBERS, AREAS, TEMPERATURE
C   CHANGES AND BODY FORCE COMPONENTS.
      WRITE(6,62) (M,NPI(M),NPJ(M),NPK(M),MATM(M),AREA(M),DELTAT(M),
     1             XBAR(M),YBAR(M),M=1,NEL)
   62 FORMAT(13HOELEMENT DATA // 72H    M     I    J     K  MAT     AREA
     1    DELTAT       XBAR           YBAR    / (1X,4I5,I3,4E12.4))
      RETURN
      END
```

Figure 6.9   Subprogram for writing out mesh data

6.6.6 The subprogram for applying the boundary conditions   The subprogram BCS shown in figure 6.10 serves to apply boundary conditions of the types discussed in section 6.5. The first action is to set all the externally applied nodal point force components to zero in preparation for any force boundary conditions. The number of nodes at which forces are prescribed, the number of distributed forces, and the number of nodes at which displacement conditions are prescribed are read into the variables NBC1P, NBC2F and NBC3P respectively. Any applied nodal point force components are then read directly into the arrays FX and FY. If there are any distributed external forces, then for each such loading in turn the number of nodes over which the force is distributed and its components in the co-ordinate directions are read into NBP, PX and PY respectively. These components are forces per unit surface area, that is per unit length of domain boundary, and can be due to both imposed tensile and shear stresses. From the next card or cards the numbers assigned to the nodes involved are read into array NPB. For example, if the distributed force shown in figure 6.4b is to be applied, the values three, zero and the magnitude of σ_{yy} are read into NBP, PX and PY respectively, followed by the numerical

```
      SUBROUTINE  BCS
C
C   SUBPROGRAM TO APPLY THE BOUNDARY CONDITIONS.
C
      COMMON /CMESH/ NEL,NNP,X(121),Y(121),AI(200),AJ(200),AK(200),
     1    BI(200),BJ(200),BK(200),AREA(200),NPI(200),NPJ(200),NPK(200),
     2    NBP,NPB(40),MOUT
     3         /CEQNS/ OKXX(121,9),OKXY(121,9),OKYX(121,9),OKYY(121,9),
     4    U(121),V(121),FX(121),FY(121),FXMOD(121),FYMOD(121),SFXX(121),
     5    SFXY(121),SFYX(121),SFYY(121),NPA(121,9),NAP(121)
     6         /CREST/ NCOND(40),TANG(40),UPRES(40),VPRES(40),NBC3P
      DO 1 I=1,NNP
      FX(I)=0.
    1 FY(I)=0.
C
C   INPUT THE NUMBERS OF SETS OF DATA FOR EACH TYPE OF BOUNDARY CONDITION
      READ(5,51) NBC1P,NBC2F,NBC3P
   51 FORMAT(14I5)
C
C   INPUT AND APPLY POINT FORCE DATA.
      IF(NBC1P.EQ.0) GO TO 2
      READ(5,52) (I,FX(I),FY(I),N=1,NBC1P)
   52 FORMAT(3(I4,2F10.0))
C
C   INPUT AND APPLY DISTRIBUTED FORCE DATA.
    2 IF(NBC2F.EQ.0) GO TO 4
      DO 3 IF=1,NBC2F
      READ(5,52) NBP,PX,PY
      READ(5,51) (NPB(N),N=1,NBP)
      NS=NBP-1
      DO 3 IS=1,NS
      I1=NPB(IS)
      I2=NPB(IS+1)
      SIDE=SQRT((X(I1)-X(I2))**2+(Y(I1)-Y(I2))**2)
      FXM=0.5*PX*SIDE
      FX(I1)=FX(I1)+FXM
      FX(I2)=FX(I2)+FXM
      FYM=0.5*PY*SIDE
      FY(I1)=FY(I1)+FYM
    3 FY(I2)=FY(I2)+FYM
C
C   DEFINE FINAL MODIFIED EXTERNAL FORCES ON THE NODES.
    4 DO 5 I=1,NNP
      FXMOD(I)=FXMOD(I)+FX(I)
    5 FYMOD(I)=FYMOD(I)+FY(I)
C
C   INPUT AND APPLY THE RESTRAINED NODE DATA.
      READ(5,53) (NPB(N),NCOND(N),TANG(N),UPRES(N),VPRES(N),N=1,NBC3P)
   53 FORMAT(2(I4,I2,3F10.0))
      DO 10 N=1,NBC3P
      I=NPB(N)
      IF(NCOND(N)-1) 8,7,6
C
C   NODE RESTRAINED TO MOVE IN DIRECTION WHOSE SLOPE IS GIVEN BY TANG.
    6 SFXX(I)=(SFXX(I)*SFYY(I)-SFXY(I)*SFYX(I))/
     1         (SFXX(I)*TANG(N)**2-(SFYY(I)+SFYX(I))*TANG(N)+SFYY(I))
      SFXY(I)=SFXX(I)*TANG(N)
      SFYX(I)=SFXY(I)
      SFYY(I)=SFXY(I)*TANG(N)
      GO TO 10
C
C   NODE RESTRAINED TO MOVE IN Y-DIRECTION ONLY.
    7 SFYY(I)=SFYY(I)-SFYX(I)*SFXY(I)/SFXX(I)
      GO TO 9
C
C   NODAL POINT DISPLACEMENTS PRESCRIBED.
    8 SFYY(I)=0.
      U(I)=UPRES(N)
      V(I)=VPRES(N)
    9 SFXX(I)=0.
      SFXY(I)=0.
      SFYX(I)=0.
   10 CONTINUE
      RETURN
      END
```

Figure 6.10   Subprogram for applying the boundary conditions

values of q, r and s into NPB. The distributed force is replaced by the equivalent nodal point forces by the method outlined in section 6.5.1. The variable NS is used to store the number of element sides over which the force is distributed, I1 and I2 store the numbers assigned to the two nodes associated with a particular side, SIDE stores its length, and FXM and FYM store the components of the equivalent forces acting at the nodes. These components are added to the relevant overall components stored in arrays FX and FY. Having applied all the force boundary conditions, the external force components are added to arrays FXMOD and FYMOD which already contain the modifications for both body forces and thermal effects.

The data for the restrained nodes at which displacement conditions are prescribed are read in the form of node number, condition number, tangent of the angle of slope of the direction of unrestrained motion (the parameter $\tan \gamma$ used in section 6.5.2) and the prescribed values of the node displacements. These are stored in the arrays NPB, NCOND, TANG, UPRES and VPRES respectively, the last four of which are located in the COMMON block of storage named CREST in order to make the data accessible to subprogram OUTPUT. The values read into NCOND are either zero, one or two according to whether the displacements of the corresponding node are prescribed, confined to the y-direction or confined to the direction whose tangent of slope is read into TANG. The values read into UPRES and VPRES are only used when the condition number is zero, and the value read into TANG only when the number is two. The self-flexibility submatrix coefficients for the restrained nodes are modified as indicated in either equations 6.40, 6.48 and 6.49, or 6.45 to 6.47, according to the condition number.

There is considerable scope for improving subprogram BCS in terms of testing for unacceptable data. For example, in the present version no provision is made for testing whether the numbers of sets of data read into the various arrays exceed their dimensions. Similarly, the nodal point numbers read in might be invalid for the particular mesh.

6.6.7 The Gauss—Seidel subprogram for biharmonic problems   The subprogram SOLVE2 shown in figure 6.11 is very similar to SOLVE1 described in section 3.8.3 for harmonic problems. The main difference is that single coefficients appearing in equation 3.72 and SOLVE1 become submatrices or subvectors in equation 6.34 and SOLVE2. Also, instead of dividing by the self-stiffness coefficients to obtain the changes in the unknowns, the previously computed self-flexibility coefficients are used as multipliers. Note that the value of the subprogram argument NNP is assigned to the local variable NPEQN, which serves to define the number of pairs of equations to be solved. Similarly, row and column numbers IROW, ICOL and IC refer to rows and columns of submatrices and subvectors as in the main program.

```
      SUBROUTINE  SOLVE2(NNP)
C
C  SUBPROGRAM FOR SOLVING BY GAUSS-SEIDEL METHOD THE LINEAR EQUATIONS
C  OBTAINED FROM THE FINITE ELEMENT FORMULATION OF BIHARMONIC PROBLEMS.
C
      COMMON /CEQNS/ OKXX(121,9),OKXY(121,9),OKYX(121,9),OKYY(121,9),
     1  U(121),V(121),FX(121),FY(121),FXMOD(121),FYMOD(121),SFXX(121),
     2  SFXY(121),SFYX(121),SFYY(121),NPA(121,9),NAP(121)
      NPEQN=NNP
C
C  INPUT THE SOLUTION PARAMETERS.
      READ(5,51) NCYCLE,IFREQ,ORELAX,TOLER
   51 FORMAT(2I5,2F10.0)
      WRITE(6,61) ORELAX
   61 FORMAT(48HOSOLUTION OF EQUATIONS BY GAUSS-SEIDEL ITERATION //
     1          25H OVER-RELAXATION FACTOR =,F6.3)
C
C  SET UP ITERATION LOOP.
      IF(IFREQ.NE.0) WRITE(6,62)
   62 FORMAT(21H  ITER        ERROR    )
      DO 3 ITER=1,NCYCLE
      SUMD=0.
      SUMDD=0.
C
C  OBTAIN NEW ESTIMATE FOR EACH UNKNOWN IN TURN.
      DO 2 IROW=1,NPEQN
      IF(SFXX(IROW)+SFYY(IROW).EQ.0) GO TO 2
      SUMX=FXMOD(IROW)
      SUMY=FYMOD(IROW)
      ICMAX=NAP(IROW)
      DO 1 IC=1,ICMAX
      ICOL=NPA(IROW,IC)
      SUMX=SUMX-OKXX(IROW,IC)*U(ICOL)-OKXY(IROW,IC)*V(ICOL)
    1 SUMY=SUMY-OKYX(IROW,IC)*U(ICOL)-OKYY(IROW,IC)*V(ICOL)
      DELU=SFXX(IROW)*SUMX+SFXY(IROW)*SUMY
      DELV=SFYX(IROW)*SUMX+SFYY(IROW)*SUMY
      SUMDD=SUMDD+ABS(DELU)+ABS(DELV)
      U(IROW)=U(IROW)+ORELAX*DELU
      V(IROW)=V(IROW)+ORELAX*DELV
      SUMD=SUMD+ABS(U(IROW))+ABS(V(IROW))
    2 CONTINUE
C
C  TEST FOR CONVERGENCE.
      ERROR=SUMDD/SUMD
      IF(ERROR.LT.TOLER) GO TO 4
C
C  OUTPUT PROGRESS INFORMATION EVERY IFREQ CYCLES, UNLESS IFREQ=0.
      IF(IFREQ.EQ.0) GO TO 3
      IF(MOD(ITER,IFREQ).EQ.0) WRITE(6,63) ITER,ERROR
   63 FORMAT(1X,I5,E15.4)
    3 CONTINUE
C
C  NORMAL EXIT FROM ITERATION LOOP INDICATES FAILURE TO CONVERGE.
      WRITE(6,64) NCYCLE
   64 FORMAT(21HONO CONVERGENCE AFTER,I5,7H CYCLES)
      RETURN
C
C  OUTPUT NUMBER OF ITERATIONS AND TOLERANCE FOR CONVERGED SOLUTION.
    4 WRITE(6,65) TOLER,ITER
   65 FORMAT(38HOITERATION CONVERGED TO A TOLERANCE OF,E12.4,
     1          6H AFTER,I5,7H CYCLES)
      RETURN
      END
```

Figure 6.11   Subprogram for applying the Gauss–Seidel method

```
      SUBROUTINE  OUTPUT
C
C  SUBPROGRAM TO OUTPUT THE FINAL RESULTS.
C
      REAL  NU
      COMMON /CMESH/ NEL,NNP,X(121),Y(121),AI(200),AJ(200),AK(200),
     1      BI(200),BJ(200),BK(200),AREA(200),NPI(200),NPJ(200),NPK(200),
     2      NBP,NPB(40),MOUT
     3            /CEQNS/ OKXX(121,9),OKXY(121,9),OKYX(121,9),OKYY(121,9),
     4      U(121),V(121),FX(121),FY(121),FXMOD(121),FYMOD(121),SFXX(121),
     5      SFXY(121),SFYX(121),SFYY(121),NPA(121,9),NAP(121)
     6            /CREST/ NCOND(40),TANG(40),UPRES(40),VPRES(40),NBC3P
     7            /CMATL/ NMAT,E(5),NU(5),ALPHA(5),RHO(5),MATM(200)
     8            /CLOAD/ DELTAT(200),XBAR(200),YBAR(200)
C
C  OUTPUT THE DISPLACEMENT BOUNDARY CONDITIONS.
      WRITE(6,61)  (NPB(IB),NCOND(IB),TANG(IB),IB=1,NBC3P)
   61 FORMAT(33HODISPLACEMENT BOUNDARY CONDITIONS //
     1       60H NODE  COND    TANG  NODE  COND     TANG  NODE  COND
     2TANG / (3(1X,I4,I5,F10.4)))
C
C  OUTPUT THE NODAL POINT FORCES AND DISPLACEMENTS.
      WRITE(6,62)  (I,FX(I),FY(I),FXMOD(I),FYMOD(I),U(I),V(I),I=1,NNP)
   62 FORMAT(37HONODAL POINT FORCES AND DISPLACEMENTS //
     1       78H  NODE   FX        FY           FXMOD      FYMOD
     2   U         V     / (1X,I5,6E12.4))
C
C  COMPUTE AND OUTPUT THE ELEMENT STRAIN AND STRESS COMPONENTS.
      WRITE(6,63)
   63 FORMAT(90HO    M   EXX         EYY          EXY         ET
     1  SIGXX       SIGYY       SIGXY   )
      DO 1 M=1,NEL
      I=NPI(M)
      J=NPJ(M)
      K=NPK(M)
      EXX=0.5*(BI(M)*U(I)+BJ(M)*U(J)+BK(M)*U(K))/AREA(M)
      EYY=0.5*(AI(M)*V(I)+AJ(M)*V(J)+AK(M)*V(K))/AREA(M)
      EXY=0.5*(AI(M)*U(I)+BI(M)*V(I)+AJ(M)*U(J)+BJ(M)*V(J)+AK(M)*U(K)
     1       +BK(M)*V(K))/AREA(M)
      MAT=MATM(M)
      ET=ALPHA(MAT)*DELTAT(M)
      FACT=E(MAT)/(1.-NU(MAT)**2)
      SIGXX=FACT*((EXX-ET)+NU(MAT)*(EYY-ET))
      SIGYY=FACT*(NU(MAT)*(EXX-ET)+(EYY-ET))
      SIGXY=FACT*0.5*(1.-NU(MAT))*EXY
    1 WRITE(6,64) M,EXX,EYY,EXY,ET,SIGXX,SIGYY,SIGXY
   64 FORMAT(1X,I5,7E12.4)
      RETURN
      END
```

Figure 6.12   Subprogram for calculating strains and stresses and writing out the results

6.6.8 The subprogram for presenting the results   The subprogram OUTPUT shown in figure 6.12 serves to write out a typical quantity of results from the finite element analysis. The displacement boundary conditions are first written out, mainly to facilitate the checking of the input data. Then the nodal point displacements, which are the primary results of the analysis, are written out together with both the modified and unmodified external forces applied to the nodes, again for checking purposes.

The secondary results presented by the present version of the subprogram are the element strains and stresses e, e_T and σ, the components of which are computed with aid of equations 6.4, 6.12 and 6.11 and stored in variables EXX,

EYY, EXY, ET, SIGXX, SIGYY and SIGXY respectively. The normal components e_{zz} and σ_{zz}, one of which is zero, could also be computed and written out. In some problems, the stresses acting at the nodal points provide useful additions to the element stress data. These can be obtained by an averaging process. For example, the stress components at a particular node could be computed as the simple averages of the values of the components associated with the elements surrounding the node. Further results include principal stresses and their directions for both elements and nodes, and if graph plotting facilities are available further coding can be written to draw the mesh and the displacement and stress profiles.

7 Some Biharmonic Problems

The case studies described in this chapter provide practical examples of the application of the finite element analysis and computer program described in chapter 6 to problems of the biharmonic plane strain and plane stress types outlined in chapter 2. Three problems are considered: a straightforward example of plane strain compression, an example concerning the stresses induced by raising the temperature of a pair of concentric cylinders, and a relatively sophisticated investigation of the stress concentration near a small hole in a flat plate under uniform tension. The only part of the program not tested by these problems is that concerned with the application of body forces. Detailed discussions of the effects of element size, shape and distribution, and the convergence and relative efficiency of the Gauss–Seidel method are confined to the last case study.

7.1 Case Study: Plane Strain Compression

Figure 7.1a shows the square cross-section of a long bar compressed between flat parallel platens. The deformation may be assumed to occur under the conditions of plane strain discussed in section 2.2.5. Since the cross-section is symmetrical about both the horizontal and vertical centre lines, only one quadrant such as the one shaded in figure 7.1a need be considered in order to compute the displacements and stresses due to the compression. This quadrant is shown in figure 7.1b, together with the global co-ordinates for the problem. The results are affected by the elastic properties of both the bar and the platens, and the amount of friction between them.

7.1.1 Problem specification
The change in shape and the distribution of vertical compressive stress along the horizontal centre line of the cross-section of a square bar with sides of unit length are to be computed. First, results are to be obtained for a uniform applied compressive stress equal in magnitude to 0.1 per cent of the Young's modulus of the bar material, on the assumption that there is no friction between the bar and the platens. Second, the friction is to be

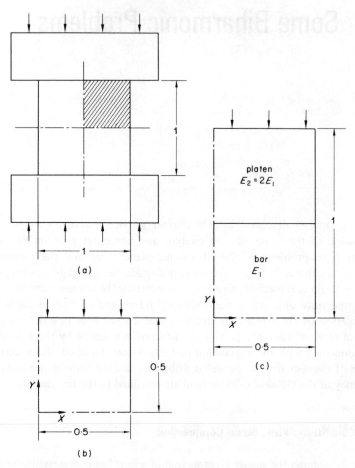

Figure 7.1   Square bar subjected to plane strain compression: (a) cross-section of bar and platens; (b) one quadrant of the bar cross-section; (c) one quadrant of the combined bar and platen

assumed sufficient to prevent slip and results are to be obtained for a 0.1 per cent reduction in the distance between the platens which are assumed to be perfectly rigid. The minimum value of the coefficient of friction necessary to prevent slip is to be estimated. Finally, for the same overall applied displacement, the effects of using flexible platens are to be examined by taking the thickness of each to be half that of the bar and Young's modulus of the platen material to be twice that of the bar. The solution domain shown in figure 7.1c is to be used to analyse one quadrant of this combined bar and platen problem.

Meshes of mainly isosceles triangular finite elements of the type shown in figure 4.5 are to be used, containing either 7 x 7 nodal points along the co-ordinate axes (52 nodes and 78 elements) for the bar alone, or 15 x 7 nodal points (97 nodes and 156 elements) for the combined bar and platen. Poisson's ratio for both materials is 0.35.

7.1.2 Solution

A computer program for solving this problem is described in section 6.6. Figure 6.5 shows the main program and figures 6.6 to 6.12 show the subprograms required with the exception of those concerned with the provision and modification of mesh data. For the present problem the version of subprogram MESH shown in figure 4.6 is needed, together with the form of MODIFY shown in figure 5.1 for applying linear scaling to the co-ordinates of the nodal points.

The data required may be listed in order as follows.

(1) Problem title and type of case required by the main program (figure 6.5).

(2) The numbers of nodal points along the co-ordinate axes and a value of the mesh data output control parameter read into MESH (figure 4.6).

(3) The overall mesh depth and width (0.5 and 0.5 for the bar alone, 1.0 and 0.5 for the combined bar and platen) read into MODIFY (figure 5.1).

(4) The material properties (including $E_1 = 1$ for the bar and $E_2 = 2$ for the platen, and zero values for the densities and coefficients of thermal expansion) read into MATLS (figure 6.6).

(5) A zero value for the temperature change read into TEMPS (figure 6.7).

(6) The boundary conditions read into BCS (figure 6.10).

(7) The maximum number of cycles of iteration, output frequency, over-relaxation factor and convergence tolerance required by SOLVE2 (figure 6.11).

The boundary conditions for the solution domains shown in figures 7.1b and c include zero prescribed displacement components for node 1 at the bottom left-hand corner in each case. The other nodes along the bottom boundaries are free to move in the horizontal direction while those on the left-hand sides are free to move vertically. The possibility of rigid body motion discussed in section 6.5.2 is thereby eliminated. The applied stress case is accommodated by prescribing a uniformly distributed vertical force of −0.001 (equivalent to a stress equal in magnitude to 0.1 per cent of E_1) along the top boundary. For the applied displacement cases, vertical displacements of −0.0005 (half of the overall displacement equal to 0.1 per cent of the distance between the platens) and zero horizontal displacements are prescribed for the nodes along the top boundaries of the domains.

For the combined bar and platen problem, subprogram MATLS shown in figure 6.6 is modified to assign the material numbers one and two to elements in the bottom and top halves of the mesh respectively. A simple way of doing this is with the aid of a calculation for the Y co-ordinate of the centroid of each element to determine its position.

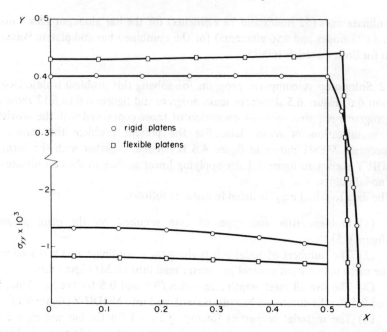

Figure 7.2   Results for rigid and flexible platens with no slip between the bar and platens

7.1.3 Results   The computed results for uniform applied stress with no friction at the platens show the expected uniform stresses and strains for pure plane strain compression. The values obtained are accurate to at least the number of significant figures printed out because meshes of CST elements are capable of representing uniform strain fields exactly. This example provides a useful, though by no means exhaustive, test for the program.

The results obtained for applied displacements and no slip are plotted in figure 7.2 for the cases of both rigid and flexible platens. The nodal point displacements are shown magnified 200 times to exaggerate the comparison of the deformed shapes of the bar quadrant with its original square form. The distributions of σ_{yy} along the bottom boundaries of the solution domains are plotted with the aid of computed element stresses shown at the original horizontal positions of the centroids. The expected 'barrelling' of the bar cross-section is obtained, particularly when the platen is rigid, and the distributions of compressive stress are by no means uniform. The effects of using flexible platens are considerable, due to the relative dimensions and Young's moduli of the bar and platens. In a more thorough investigation the effects of varying both the widths and thicknesses of the platens would need to be considered.

The minimum value of the coefficient of friction necessary to prevent slip at a rigid platen may be obtained by examining the σ_{yy} and σ_{xy} stress components computed for the elements forming the top boundary of the solution domain. The shear stress is very small near the vertical axis of symmetry of the bar cross-section, but rises to about 28 per cent of the corresponding value of σ_{yy} near the outer edge. Consequently, a coefficient of friction of at least 0.28 is required to prevent slip.

7.2   Case Study: Stresses in Concentric Cylinders

Figure 7.3 shows a cross-sectional view of two long concentric cylinders made from different materials. At ambient temperature they fit together exactly and have outer radii of r_1 and r_2 as shown. Let Young's modulus, Poisson's ratio and coefficient of thermal expansion be E_1, ν_1 and α_1 for the material of the inner solid cylinder, and E_2, ν_2 and α_2 for that of the outer hollow one. If $\alpha_1 > \alpha_2$ and the temperature of the system is raised, strains and stresses are induced in both cylinders. The mode of deformation may be assumed to be of the plane strain type.

7.2.1 Problem specification   The stresses induced by a uniform temperature rise are to be computed and compared with the analytical solution for this problem. A circular mesh of finite elements of the type shown in figure 4.9 is to be used. The value of α_1 is 2.0×10^{-5} per unit temperature difference, which is twice that of α_2. Poisson's ratios of the two materials are both 0.32, but their Young's moduli are such that $E_2 = 2E_1$. The ratio of the cylinder radii is $r_2/r_1 = 2.5$, and the temperature rise is one of 100 temperature units.

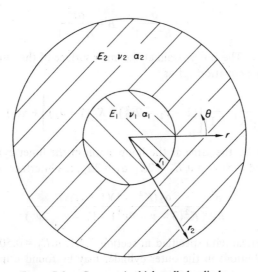

Figure 7.3   Concentric thick-walled cylinders

7.2.2 Analytical solution   For the purposes of analysis it is convenient to use the cylindrical polar co-ordinates r and θ shown in figure 7.3, together with an axial co-ordinate z, normal to the cross-section. Since the problem is axi-symmetric the only nonzero stress components are the radial, hoop and axial direct stresses σ_{rr}, $\sigma_{\theta\theta}$ and σ_{zz}. Under conditions of plane strain, the hoop strain in either cylinder is related to the stresses by an equation of the form of, say, the first of equations 6.15

$$e_{\theta\theta} = \frac{1}{E}\left[(1 - \nu^2)\,\sigma_{\theta\theta} - \nu(1 + \nu)\,\sigma_{rr}\right] + \alpha(1 + \nu)\,\Delta T \qquad (7.1)$$

Let σ be the magnitude of the compressive stress at the interface between the two cylinders. Hence, the radial and hoop stresses in the inner cylinder are

$$\sigma_{rr} = \sigma_{\theta\theta} = -\sigma \qquad (7.2)$$

and the hoop strain may be obtained from equation 7.1 as

$$e_{\theta\theta} = -\frac{\sigma}{E_1}(1 - \nu_1 - 2\nu_1^2) + \alpha_1(1 + \nu_1)\,\Delta T \qquad (7.3)$$

Using the Lamé equations (see, for example, Ford (1963)), the distributions of radial and hoop stresses in the outer cylinder may be expressed as

$$\sigma_{rr} = A - \frac{B}{r^2}, \quad \sigma_{\theta\theta} = A + \frac{B}{r^2} \qquad (7.4)$$

where the constants A and B can be found from the boundary conditions $\sigma_{rr} = -\sigma$ at $r = r_1$ and $\sigma_{rr} = 0$ at $r = r_2$ as

$$A = \frac{\sigma}{K^2 - 1}, \quad B = \frac{\sigma r_2^2}{K^2 - 1} \qquad (7.5)$$

where $K = r_2/r_1$. The hoop strain at the inner surface of the outer cylinder may be obtained from equation 7.1 as

$$e_{\theta\theta} = \frac{\sigma}{E_2(K^2 - 1)}\left[(1 - \nu_2 - 2\nu_2^2) + K^2(1 + \nu_2)\right] + \alpha_2(1 + \nu_2)\,\Delta T \quad (7.6)$$

Since this must be the same as the hoop strain in the inner cylinder given by equation 7.3, and $\nu_1 = \nu_2 = \nu$, $E_2 = 2E_1$, $\alpha_1 = 2\alpha_2$, the interface stress is given by

$$\frac{\sigma}{E_2} = \frac{\alpha_2(K^2 - 1)(1 + \nu)\,\Delta T}{K^2(3 - \nu - 4\nu^2) - (1 - \nu - 2\nu^2)} \qquad (7.7)$$

Using the numerical data specified in section 7.2.1, $\sigma/E_2 = 0.5053 \times 10^{-3}$, and the stress distributions in the outer cylinder may be found using equations 7.4 and 7.5.

7.2.3 Finite element solution and results   The stresses in the cylinders may be computed using the set of finite element subprograms described in section 7.1.2, with the exception that the version of MESH shown in figure 4.10 is employed. It is convenient to have six elements at the centre of the mesh and six nodal points along a horizontal radius, which, as explained in section 4.3.4, result in totals of 91 nodes and 150 elements. This is the finest mesh of the circular type permitted by the dimensions of the arrays in the present version of the program. It is also appropriate for this problem in that the interface between the two cylinders coincides with a ring of nodes. As it is convenient to let the outer radius r_2 have the value one, there is no need to modify the basic mesh. It is necessary, however, to modify subprogram MATLS shown in figure 6.6 to assign the relevant material numbers to the elements forming the inner and outer cylinders. A simple way of doing this is with the aid of a calculation for the distance of centroid of each element from the centre of the mesh to determine its position.

The numerical data specified in section 7.2.2 are used, the value one being appropriate for E_2. Zero values are used for the material densities in order to exclude body force effects. Neither prescribed forces nor displacement restraints are applicable to the nodes on the mesh boundary. As indicated in section 6.5.2, however, it is essential to preclude rigid body motion. A convenient way to do this is by prescribing zero displacements for the centre node of the mesh and freedom to move in the horizontal direction only for the node at $r = r_2$, $\theta = 0$ (figure 7.3).

The computed displacements and stresses are axi-symmetric to within the accuracy of the results. For the elements forming the inner cylinder the direct stresses σ_{xx} and σ_{yy} are equal and constant to within the same accuracy. The average value is -0.57×10^{-3}, the magnitude of which is about 13 per cent higher than the true value of σ obtained from equation 7.7 (with $E_2 = 1$). Figure 7.4 shows the computed and true radial distributions of radial and hoop stresses. The computed values are obtained as σ_{xx} and σ_{yy} respectively for elements adjacent to the line $\theta = 0$. The points are plotted at radii corresponding to the centroids. The computed radial stresses show reasonably good agreement with the analytical solution, but the hoop stresses are less satisfactory. With only five rings of elements, the mesh is too coarse to be able to cope adequately with the abrupt change in hoop stress at the interface between the cylinders and the rapid variation near this interface in the outer cylinder.

The utilisation of the relatively large number of elements is poor. A much more accurate solution to this axi-symmetric problem using a similar number of elements could be obtained by considering only one sector of the domain. For example, a triangular mesh of the type shown in figure 4.7 could be modified to fit a $60°$ sector. An even more efficient method for solving axi-symmetric problems is outlined in section 8.1. Once axial symmetry is lost the entire circular domain must be considered, and in order to achieve acceptable accuracy considerably larger numbers of nodes and elements may be necessary.

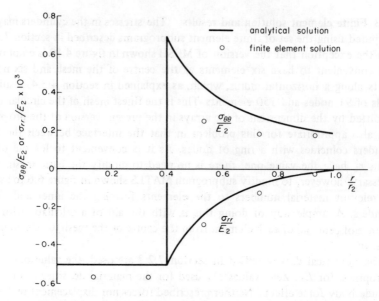

Figure 7.4   Results for stress distributions in concentric cylinders

This case study demonstrates how a finite element method formulated in terms of cartesian co-ordinates can be applied to a problem normally analysed using a quite different system. Both this problem and the one described in section 7.1 also provide examples of the application of the method to inhomogeneous solution domains.

7.3   Case Study: Stress Concentration near a Hole in a Flat Plate

Figure 7.5a shows a square flat plate subjected to a uniform tensile stress σ, with a small circular hole at its centre. The thickness of the plate may be assumed to be sufficiently small for the plane stress approximation discussed in section 2.2.6 to be applicable. Let the diameter of the hole be $2a$ and the width of the plate be $2b$. Also, let the global co-ordinates X and Y be defined for use in the subsequent finite element analysis, with the origin at the centre of the hole. The presence of the hole causes local concentrations of stresses, the greatest being at $X = \pm a$, $Y = 0$.

An analytical solution for this problem is available (see, for example, Ford (1963)) which yields the following expressions for stresses along the line $Y = 0$.

$$\sigma_{xx} = \frac{3\sigma}{2} \left(\frac{a^2}{X^2} - \frac{a^4}{X^4} \right) \tag{7.8}$$

$$\sigma_{yy} = \frac{\sigma}{2}\left(2 + \frac{a^2}{X^2} + \frac{3a^4}{X^4}\right) \tag{7.9}$$

$$\sigma_{xy} = 0 \tag{7.10}$$

and the maximum stress is $\sigma_{yy} = 3\sigma$. Since this analytical solution is derived for a plate of infinite width, its application to one of finite width involves some error. For example, the true boundary condition at $X = \pm b$ is $\sigma_{xx} = 0$, but equation 7.8 does not give this result exactly. Provided the hole is relatively small, however, the error is small. Taking the ratio $a/b = 0.05$ specified below, the value of σ_{xx} at the edge of the plate is less than 0.4 per cent of σ. The errors involved in the analytical solution are negligible compared with those associated with the finite element solution for this problem.

Although the magnitudes of the stresses near the hole are large, they diminish rapidly with distance from the hole to the values they would have in its absence. This problem therefore provides a severe test of any numerical method of solution and is a good one to illustrate the capabilities of the finite element method. It also involves a combination of rectangular and circular boundary shapes which is difficult to fit using other methods, such as those of the finite difference type.

7.3.1 Problem specification The stresses near a hole in a square flat plate having relative dimensions $a/b = 0.05$ are to be computed and compared with the analytical solution. Advantage is to be taken of the symmetry of the problem to consider only the quadrant shaded in figure 7.5a as the solution domain shown in figure 7.5b. A method of mesh data generation and modification is to be devised to concentrate relatively small elements near the edge of the hole. The effects of varying the degree of this concentration are to be investigated.

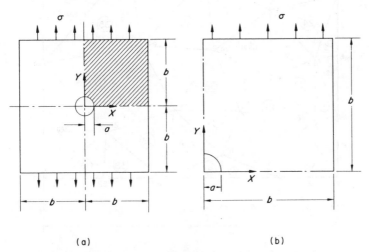

(a)   (b)

Figure 7.5   Flat plate subjected to uniform tension with a hole at its centre: (a) the entire plate; (b) one quadrant of the plate

7.3.2 Mesh data generation and modification   Since relatively small elements are required near the arc at the corner of the solution domain shown in figure 7.5b it is convenient to use a mesh with most of the nodal points arranged in rows forming concentric arcs. A similar number of nodes may be used in each row and the radial distances between successive rows progressively decreased towards the centre. Figure 7.6 shows a mesh of this type, though not drawn to the scale of the present problem, and figure 7.8 shows a small part of one of the more refined meshes actually used.

The relative positions of most of the nodes and elements in the mesh shown in figure 7.6 (specifically, nodes 1 to 27 and elements 1 to 36) are the same as in the square mesh of mainly isosceles triangular elements shown in figure 4.5. Only the co-ordinates of the nodal points are modified to obtain the required boundary shape and element distribution, and a few extra nodes and elements are added to form the corner of the plate remote from the hole. Horizontal rows of nodal points in the basic square mesh are modified to form arcs. The points adjacent to the ends of the outermost arc (nodes 24 and 26 in figure 7.6) are

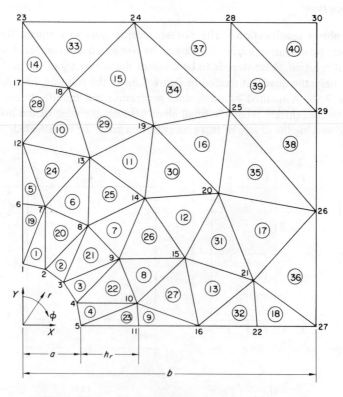

Figure 7.6   A mesh for the stress concentration problem

then moved either vertically or horizontally to the edges of the domain to avoid the creation of long thin elements at the top left and bottom right-hand corners of the mesh. The extra nodal points on the top boundary of the mesh have the same X co-ordinates as the corresponding nodes on the outermost arc, while those on the right-hand boundary have the same Y co-ordinates. The last nodal point is at the top right-hand corner of the mesh. This modification process can be applied to a square mesh containing any numbers of points n_x and n_y in the co-ordinate directions, although for ease of subsequent analysis and programming it is convenient to assume that both are odd.

Since elements are to be concentrated near the edge of the hole and the degree of this concentration is to be varied, it is convenient to introduce a scale factor S to define a constant ratio of radial distances between successive rows of nodes. Let h_r be the distance between the first two rows as shown in figure 7.6. In general there are n_y such rows and

$$h_r (1 + S + S^2 + \ldots + S^{n_y - 2}) = b - a$$

$$h_r = \frac{(b - a)(S - 1)}{S^{n_y - 1} - 1} \quad \text{for } S \neq 1 \qquad (7.11)$$

The Y co-ordinates of the nodal points in the basic mesh can be first modified with the aid of this result to the form

$$Y_i^* = \frac{h_r(S^{i_y - 1} - 1)}{S - 1} \qquad (7.12)$$

where i_y is the number of the row in which node i occurs. Then the required curvature can be introduced by a second modification. If the polar co-ordinates r and ϕ shown in figure 7.6 are used, the modified position of the typical node i is given by

$$r = a + Y_i^*, \quad \phi = \tfrac{1}{2} \pi X_i \qquad (7.13)$$

and the final global co-ordinates are

$$X_i^{**} = r \sin \phi, \quad Y_i^{**} = r \cos \phi \qquad (7.14)$$

Let i_1 and i_2 be the numbers assigned to the nodal points adjacent to the ends of the outermost row (nodes 24 and 26 in figure 7.6). Their values are obtained by subtracting $n_x - 2$ and 1 respectively from the total number of nodes in the basic mesh. The co-ordinates of these two nodal points are further modified as follows

$$Y_{i_1}^{***} = b, \quad X_{i_2}^{***} = b \qquad (7.15)$$

The additional row of nodal points (nodes 28 and 29 in figure 7.6) that excludes the last corner point involves in general $n_x - 3$ points, where n_x is odd. If i_x is used to count points along this row, then their co-ordinates are given by

$$X_i = X_{i_1 + i_x}, \quad Y_i = b, \quad \text{for } i_x \leqslant \tfrac{1}{2}(n_x - 3)$$
$$X_i = b, \quad Y_i = Y_{i_1 + i_x - 1}, \quad \text{for } i_x > \tfrac{1}{2}(n_x - 3) \tag{7.16}$$

The total number of nodes is found by adding $n_x - 2$ to the total for the basic mesh, and the co-ordinates of the last point are both equal to b.

In order to define the node numbers of the extra elements it is convenient to consider first the outward pointing elements with the exception of the outermost one (elements 37 and 38 in figure 7.6). The numbers of the nodes of a typical element of this type numbered m are

$$i = i_1 + m - m_1 - 1, \quad j = i + 1, \quad k = i + n_x - 1 \tag{7.17}$$

where m_1 is the total number of elements in the basic mesh. Let m_2 be the number of such elements plus the extra outward pointing ones just numbered. The numbers of the nodes of a typical extra inward pointing element (element 39 in figure 7.6) numbered m may therefore be defined as

$$i = i_1 + m - m_2, \quad j = i + n_x - 1, \quad k = j - 1 \tag{7.18}$$

The total number of elements is found by adding $2n_x - 6$ to the total for the basic mesh, and the numbers of the nodes of the last one are

$$i = i_2 + \tfrac{1}{2}(n_x - 3) + 1, \quad j = i + 1, \quad k = i_2 + n_x - 1 \tag{7.19}$$

Figure 7.7 shows a version of subprogram MODIFY for modifying the mesh data generated by the version of MESH shown in figure 4.6 according to the above method. The variables NXPT, NYPT, IX and IY are used to store the values of n_x, n_y, i_x and i_y respectively, while I, I1, I2, M and M1 serve to store the values of i, i_1, i_2, m and m_1. IXMAX stores the maximum number of nodes or elements in a particular row, and YIMOD stores the Y co-ordinate modified according to equation 7.12 for a particular row of nodes. The values of the mesh modification parameters S, a and b are stored in variables S, A and B, while those of h_r, the polar co-ordinates r and ϕ, and π are stored in HR, R, PHI and PI.

7.3.3 Finite element solution and results
The stress distributions in the plate may be computed using the set of finite element subprograms described in section 7.1.2, with the exception that the versions of MESH and MODIFY shown in figures 4.6 and 7.7 are employed. It is desirable to use a rather finer mesh than that shown in figure 7.6, and a basic mesh with $n_x = n_y = 9$ giving totals of 92 nodal points and 148 elements in the modified mesh is appropriate. This is the finest mesh of the required form acceptable to the present version of the program. The numerical data supplied to the program include values of 0.05 and 1.0 for a and b, any reasonable values such as 1.0 and 0.3 for Young's modulus and Poisson's ratio, together with zero values for the material density

```
      SUBROUTINE  MODIFY
C
C  SUBPROGRAM TO MODIFY A MESH TO SUIT A PARTICULAR PROBLEM.
C  THIS VERSION ADAPTS A SQUARE MESH TO STRESS CONCENTRATION PROBLEM.
C
      COMMON /CMESH/ NEL,NNP,X(121),Y(121),AI(200),AJ(200),AK(200),
     1     BI(200),BJ(200),BK(200),AREA(200),NPI(200),NPJ(200),NPK(200),
     2     NBP,NPB(40),MOUT
     3          /CMPAR/ NXPT,NYPT
C
C  INPUT THE MESH SCALE FACTOR AND THE PLATE DIMENSIONS.
      READ(5,51) S,A,B
  51  FORMAT(3F10.0)
C
C  TEST FOR ACCEPTABLE BASIC MESH.
      IF(MOD(NXPT,2).EQ.1.AND.MOD(NYPT,2).EQ.1) GO TO 1
      WRITE(6,61)
  61  FORMAT(41HOMESH UNSUITABLE FOR PRESENT MODIFICATION)
      STOP
C
C  PERFORM FIRST MODIFICATION OF Y CO-ORDINATES.
   1  HR=(B-A)*(S-1.)/(S**(NYPT-1)-1.)
      I=0
      DO 2 IY=1,NYPT
      YIMOD=HR*(S**(IY-1)-1.)/(S-1.)
      IXMAX=NXPT
      IF(MOD(IY,2).EQ.0) IXMAX=NXPT+1
      DO 2 IX=1,IXMAX
      I=I+1
   2  Y(I)=YIMOD
C
C  PERFORM SECOND MODIFICATION TO INTRODUCE CURVATURE.
      PI=4.*ATAN(1.)
      DO 3 I=1,NNP
      R=A+Y(I)
      PHI=X(I)*0.5*PI
      X(I)=R*SIN(PHI)
   3  Y(I)=R*COS(PHI)
C
C  MODIFY CO-ORDINATES OF POINTS NEXT TO THE END POINTS OF THE OUTERMOST
C  CIRCUMFERENTIAL ROW.
      I1=NNP-NXPT+2
      I2=NNP-1
      Y(I1)=B
      X(I2)=B
C
C  DEFINE AND TEST NEW TOTAL NUMBERS OF NODES AND ELEMENTS.
      I=NNP
      NNP=NNP+NXPT-2
      M=NEL
      NEL=NEL+2*NXPT-6
      IF(NNP.LE.121.AND.NEL.LE.200) GO TO 4
      WRITE(6,62) NNP,NEL
  62  FORMAT(30HOEXCESSIVE SIZE OF MESH, NNP =,I5,8H   NEL =,I5)
      STOP
C
C  DEFINE THE CO-ORDINATES OF THE ADDITIONAL NODES.
   4  IXMAX=NXPT-3
      DO 6 IX=1,IXMAX
      I=I+1
      I1=I1+IX
      IF(IX.GT.(NXPT-3)/2) GO TO 5
      X(I)=X(I1)
      Y(I)=B
      GO TO 6
   5  X(I)=B
      Y(I)=Y(I1-1)
   6  CONTINUE
      X(NNP)=B
      Y(NNP)=B
```

Figure 7.7   Subprogram for modifying a rectangular mesh of mainly isosceles elements into a form suitable for the stress concentration problem

```
C
C   DEFINE THE NODES OF THE ADDITIONAL ELEMENTS.
    M1=M
    DO 7 IX=1,IXMAX
    M=M+1
    NPI(M)=I1+M-M1-1
    NPJ(M)=NPI(M)+1
7   NPK(M)=NPI(M)+NXPT-1
    M2=M
    IXMAX=IXMAX-1
    DO 8 IX=1,IXMAX
    M=M+1
    NPI(M)=I1+M-M2
    NPJ(M)=NPI(M)+NXPT-1
8   NPK(M)=NPJ(M)-1
    NPI(NEL)=I2+(NXPT-3)/2+1
    NPJ(NEL)=NPI(NEL)+1
    NPK(NEL)=NNP
    RETURN
    END
```

Figure 7.7   Continued

and coefficient of thermal expansion. Various values of the mesh scale factor S are used as described below.

The boundary conditions for the solution domain shown in figure 7.5 include freedom for nodes on the bottom boundary to move horizontally and for those on the left-hand side to move vertically. Although the position of no node is prescribed, these restraints are sufficient to prevent rigid body motion. The uniform applied stress is accommodated by prescribing a uniformly distributed vertical force of magnitude one along the top boundary, which has the effect of setting $\sigma = 1$.

Investigations of the effects of varying the over-relaxation factor and convergence tolerance similar to those described in section 5.1.4 show the optimum value of ω to be about 1.8 (for $S = 2$) and that a tolerance of 10^{-6} reduces convergence errors to acceptable levels. The practical range for the value of the mesh scale factor is approximately $1 < S < 3$, and results are obtained for values of 1.5, 2.0, 2.5 and 3.0 ($S = 1$ is unacceptable to equation 7.11 and hence to subprogram MODIFY). Although a large value of S ensures very closely spaced rows of nodal points near the hole, they may be so close as to make elements in the innermost rows obtuse-angled. Figure 7.8 shows a small part of the mesh near the hole, drawn to scale for $S = 2$. Note how very small the inner elements are in relation to the size of the hole, the centre of which is not shown, and the tendency for them to become obtuse angled.

Table 7.1 shows the results obtained for the four values of S. The ratios of maximum to applied stress, $\hat{\sigma}_{yy}/\sigma$, are the computed stresses for element 8 shown in figure 7.8. The absolute maximum stress is to be expected at the position of node 9, and of the two elements having this point as a node element 8 has the centroid which is nearer the edge of the hole. Also tabulated are the numbers of cycles of iteration, q, required for convergence of the Gauss–Seidel process, and element aspect rations A_1 and A_2. In general the aspect ratio of an

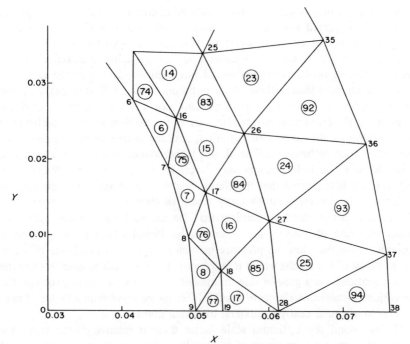

Figure 7.8   A small part of the actual mesh used for $S = 2$

element may be defined as the ratio between its largest and smallest dimensions. In the present context these dimensions may be taken as the radial distance between the successive rows of nodal points containing the particular element and the length of its side lying along one of these rows, though not necessarily in this order. A_1 and A_2 represent average values for the innermost and outermost rows of elements derived from the basic mesh, and values for the intervening rows lie between these extremes. Most of the elements in the meshes have their larger dimension in the radial direction, except perhaps those in the innermost rows. The values of A_1 marked with asterisks in table 7.1 refer to elements of this exceptional type.

Table 7.1
*Stress concentration results for various values
of the mesh scale factor*

S	$\hat{\sigma}_{yy}/\sigma$	q	A_1	A_2
1.5	2.71	63	1.6	2.0
2.0	2.77	66	2.7*	3.2
2.5	2.67	127	11*	4.1
3.0	2.61	500+	34*	4.8

*Elements with smallest dimension in the radial direction

As the value of S is increased the computed maximum stress at first improves towards the expected value of 3, and then deteriorates somewhat. At the same time the number of cycles of iteration increases, although the increase is only significant for $S > 2$. For $S = 3$ convergence is not quite achieved after 500 cycles, although further iteration would not significantly affect the tabulated maximum stress. Clearly, there is an optimum value of S at about 2. These results can be explained in terms of element aspect ratios. For values of S not exceeding 2 the maximum aspect ratios are not more than about 3, particularly in the region of the stress concentration, and the shapes of the elements are reasonably close to being equilateral. For higher values of S the aspect ratios of elements near the hole become very large, implying long thin elements. In section 6.4 it is suggested that the Gauss–Seidel method applied to biharmonic problems only converges satisfactorily when the elements are nearly equilateral. The present results tend to confirm that the use of long thin elements does reduce the rate of convergence dramatically. Perhaps even more important, however, is the fact that the presence of such elements has an adverse effect on the accuracy of the results which is independent of the method used to solve the linear equations. As a general rule, in meshes used for biharmonic problems the dimensions of adjacent elements should differ by no more than a factor of two, in order to maintain nearly equilateral triangular shapes.

Having found the optimum scale factor there is relatively little scope for further improving the accuracy of the results by modifying the distribution of elements within the particular mesh selected. Such an improvement can best be achieved by refining the mesh. Totals of 92 nodal points and 148 elements are comparatively modest for a problem of this complexity.

Because the maximum stress concentration is the most difficult to compute accurately, the values shown in table 7.1 do not provide a typical assessment of the accuracy of the present finite element method. Figure 7.9 shows the variation of σ_{yy} with distance from the hole along a small part of the X-axis very near the hole. In addition to the analytical solution given by equation 7.9, both element and nodal point stresses are plotted. Element stresses are for elements such as those numbered 77, 17 and 94 in figure 7.8 and are plotted at positions level with their centroids. Nodal point stresses are obtained by averaging as described in section 6.6.7. For example, σ_{yy} for nodes 19 and 28 in figure 7.8 are obtained as the averages of the values associated with elements 77 and 17, and 17, 85, 25 and 94 respectively. In this case nodal point stresses tend to be closer to the analytical solution than the element stresses. As is to be expected, the comparison between the computed and analytical results is worst near the position of maximum stress where the stress varies most rapidly. Similar comparisons can be made for the stress concentrations at other points on the edge of the hole. For elements remote from the hole the computed stresses are very close to the values $\sigma_{xx} = \sigma_{xy} = 0$, $\sigma_{yy} = 1$ for uniform tension.

In section 3.6.3 criteria are described for comparing the computing times **required by the** direct elimination and iterative methods of solving simultaneous

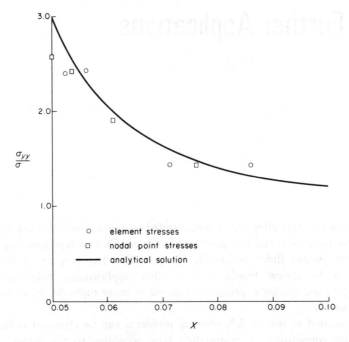

Figure 7.9   Results for stress concentration near the hole in the flat plate

linear algebraic equations. For the present biharmonic problem, the number of pairs of equations is $n = 92$, the number of nodal points; as indicated in section 6.4 comparisons between the methods can be made in terms of pairs of equations. From table 7.1 the number of cycles of Gauss–Seidel iteration is $q = 66$ (for $S = 2$). Hence, using equations 3.75 and 3.76, $r_1 = 0.21$ and $r_2 = 6.5$. These values are similar to those presented in section 5.1.4 for a typical harmonic problem. While the Gauss–Seidel method is faster than the full elimination method it is slower than elimination applied to the rectangular form of stiffness matrix. Again in view of the relative storage requirements, however, the choice between direct and iterative methods is not clearcut.

8 Further Applications

In previous chapters attention is concentrated on simple finite element methods applicable to two-dimensional problems of the equilibrium type involving either newtonian viscous fluids or linearly elastic solids. The purpose of this final chapter is to review briefly some further applications, including three-dimensional and nonlinear problems. The use of more sophisticated elements is also discussed.

As indicated in section 2.3, physical problems can be classified as being of either the equilibrium or propagation type according to the nature of the governing partial differential equation. Time-dependent problems provide the most common examples of the propagation type. Equilibrium problems include those situations where critical values of one or more parameters are required in addition to the corresponding configuration of the system. These are often referred to as eigenvalue problems, and practical examples include the buckling of structures and stability problems in general. Both propagation and eigenvalue problems can be solved by finite element methods.

8.1 Axi-symmetric Problems

With only relatively minor modifications, the finite element methods described here for both harmonic and biharmonic problems can be used to solve three-dimensional problems of the same types which are symmetrical in terms of geometry, boundary conditions and external loading about some axis. Practical examples of such problems include heat transfer and stress analysis in the walls of cylindrical pressure vessels, torsion of shafts whose circular cross-sections vary in diameter, and fluid flow along circular pipes and annuli of varying size. The problem considered in section 7.2 provides a particularly simple example in that the geometry, boundary conditions and external (thermal) loading are independent of axial position.

Axi-symmetric problems can be solved by considering a two-dimensional solution domain lying in a plane containing the axis of symmetry. Figure 8.1

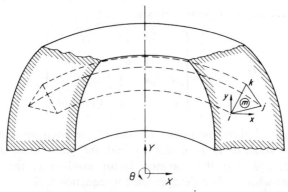

Figure 8.1   Cross-section through the axis of a body of revolution

shows part of such a domain for a typical body of revolution. In order to retain
the maximum similarity with the finite element analyses described in chapters 3
and 6, global co-ordinates X and Y are defined in the radial and axial directions
respectively. The angle θ provides the third co-ordinate in the hoop direction.
The solution domain is divided into a mesh of triangular finite elements, the
typical element being as shown in figure 3.2 except that it now represents a ring
of triangular cross-section rather than a prism of unit thickness.

Although the present approach is applicable to both harmonic and biharmonic problems it is convenient to develop the analysis with reference to, say,
the latter type involving elastic solids. Owing to the axial symmetry the only
nonzero shear stresses and strains are σ_{xy} and e_{xy}, in the plane of the solution
domain. In general, however, all the direct stress and strain components in the
radial, axial and hoop directions are nonzero. This is in contrast to plane stress
and plane strain problems where either the direct stress or strain normal to the
solution domain are zero. Using constitutive equations 2.21 to 2.24 (with hoop
components $\sigma_{\theta\theta}$ and $e_{\theta\theta}$ in place of σ_{zz} and e_{zz}), the relationships between the
stresses and strains are given by

$$
\begin{bmatrix} e_{xx} \\ e_{yy} \\ e_{\theta\theta} \\ e_{xy} \end{bmatrix} = \frac{1}{E} \begin{bmatrix} 1 & -\nu & -\nu & 0 \\ -\nu & 1 & -\nu & 0 \\ -\nu & -\nu & 1 & 0 \\ 0 & 0 & 0 & 2(1+\nu) \end{bmatrix} \begin{bmatrix} \sigma_{xx} \\ \sigma_{yy} \\ \sigma_{\theta\theta} \\ \sigma_{xy} \end{bmatrix} + \alpha\Delta T \begin{bmatrix} 1 \\ 1 \\ 1 \\ 0 \end{bmatrix} \tag{8.1}
$$

which may be inverted to give

$$
\sigma = D(e - e_T) \tag{8.2}
$$

σ, e and e_T being the vectors of stresses, strains and thermal strains displayed in equation 8.1. The elastic property matrix is

$$D = \frac{E^*}{1 - \nu^{*2}} \begin{bmatrix} 1 & \nu^* & \nu^* & 0 \\ \nu^* & 1 & \nu^* & 0 \\ \nu^* & \nu^* & 1 & 0 \\ 0 & 0 & 0 & \frac{1}{2}(1 - \nu^*) \end{bmatrix} \tag{8.3}$$

where the modified elastic properties E^* and ν^* are given by equations 6.16. Note that the form of D is similar to that for plane strain displayed in equations 6.12, and that the common factor involved in the coefficients becomes infinite when $\nu = \frac{1}{2}$ as demonstrated by equation 6.17.

The hoop strain component is defined as

$$e_{\theta\theta} = \frac{u}{r} \tag{8.4}$$

where r is the radial distance from the axis of symmetry, and $r \equiv X$ in figure 8.1. This strain is not constant over individual elements, irrespective of the form of shape function used for the displacements. Because the direct equilibrium formulation used in section 6.1 is unsuitable for elements other than those of a constant strain type, the variational formulation described in section 6.3 is employed to develop the analysis. Instead of equation 6.30, the change in total potential energy for small changes in the displacements is found by integrating over the entire three-dimensional solution domain

$$d\chi = \iint 2\pi r (\sigma^T \, de) \, dx \, dy - \iint 2\pi r (\bar{X} du + \bar{Y} dv) \, dx \, dy - F^T d\delta \tag{8.5}$$

In order to evaluate this expression exactly some relatively complicated analytical or numerical integrations must be performed over the areas of the elements. If linear shape functions for the displacements are to be used it is reasonable to assume that r is constant for a particular element and equal to the radius of its centroid

$$\bar{r}_m = \frac{1}{3}(r_i + r_j + r_k) \tag{8.6}$$

It is also reasonable to assume that the hoop strain defined by equation 8.4 is constant for the element

$$e_{\theta\theta} = \frac{\bar{u}}{\bar{r}_m}, \quad \bar{u} = \frac{1}{3}(u_i + u_j + u_k) \tag{8.7}$$

The errors introduced by these assumptions are generally no more serious than those due to the use of linear shape functions. Since the elements are conforming, equation 8.5 can be expressed in terms of element matrices as

$$d\chi = 2\pi\Sigma \, \bar{r}_m \Delta_m \sigma^T de - 2\pi\Sigma \, \bar{r}_m G_m^T d\delta_m - F^T d\delta \tag{8.8}$$

The strains are related to displacements by a modified form of equation 6.4

$$e = \frac{1}{2\Delta_m} B \delta_m \qquad (8.9)$$

where the dimension matrix is

$$B = \begin{bmatrix} b_i & 0 & b_j & 0 & b_k & 0 \\ 0 & a_i & 0 & a_j & 0 & a_k \\ \dfrac{2\Delta_m}{3\bar{r}_m} & 0 & \dfrac{2\Delta_m}{3\bar{r}_m} & 0 & \dfrac{2\Delta_m}{3\bar{r}_m} & 0 \\ a_i & b_i & a_j & b_j & a_k & b_k \end{bmatrix} \qquad (8.10)$$

The final result similar to equation 6.32 is

$$2\pi \sum \frac{\bar{r}_m}{4\Delta_m} B^T D B \delta_m = F + 2\pi \sum \bar{r}_m G_m + 2\pi \sum \frac{\bar{r}_m}{2} B^T D e_T \qquad (8.11)$$

or

$$K\delta = F^* \qquad (8.12)$$

in the usual general form. Note that, in order to comply with the requirement for axi-symmetric external loading, the forces F must be uniformly distributed around the circumferences through the nodes at which they are applied.

The computer program described in section 6.6 can be readily adapted for solving axi-symmetric biharmonic problems. The main modifications necessary are the introduction of mean radii for the elements and the use of four stress and strain components as outlined in the above analysis.

8.2 Higher-order Elements

Constant strain triangular elements involving linear variations of the displacements or velocities are the simplest type available for solving two-dimensional problems. Elements involving higher-order shape functions can also be used and have both some advantages and some disadvantages. In principle, a shape function can take any mathematical form which satisfies the compatibility requirements discussed in section 2.1.4. In practice, polynomials are by far the most commonly used for most types of problems, and automatically satisfy compatibility within the elements. Similar considerations apply to problems formulated in terms of variables other than displacements or velocities, such as the torsion problems solved in section 5.2.

Suppose the linear shape function defined by, for example, equation 3.5 is replaced by the quadratic form

$$w(x, y) = C_1 + C_2 x + C_3 y + C_4 x^2 + C_5 y^2 + C_6 xy \qquad (8.13)$$

The constants C_1 to C_6 have to be obtained from six values of w, or, in general, of w and its derivatives, at nodal points associated with the element concerned. Figure 8.2a shows a suitable triangular element having nodes at the centres of its sides in addition to those at the corners: the constants may be obtained in terms of the six nodal point values of w. Sections 8.4 and 8.5 provide examples of the use of derivatives as additional nodal point variables. The strains or strain rates obtained from a quadratic shape function are linear functions of position, and the element shown in figure 8.2a is therefore known as a linear strain triangle. Similarly, the element shown in figure 8.2b is a quadratic strain triangle, having the following cubic shape function

$$w(x,y) = C_1 + C_2 x + C_3 y + C_4 x^2 + C_5 y^2 + C_6 xy$$
$$+ C_7 x^3 + C_8 y^3 + C_9 x^2 y + C_{10} xy^2 \tag{8.14}$$

The ten coefficients associated with the full cubic polynomial can be accommodated by using ten nodes, four per side of the element and one at the centroid, as shown. Alternatively, the node at the centroid could be eliminated and the number of independent parameters in the shape function reduced to nine by setting, say, $C_9 = C_{10}$.

While there is no necessity for triangular elements to be used, they are particularly well suited both to fitting irregular boundary shapes and to meeting the requirements of the various orders of shape functions in terms of numbers of nodal points. For example, a quadrilateral element in its simplest form would have four nodes: one more than the number required for a linear shape function, but two less than for a quadratic one.

As indicated in section 3.7.1, it is desirable for any finite element to be of the conforming type. For the analyses described in chapters 3 and 6 this requirement is satisfied if the displacements or velocities are continuous across the boundaries between the elements. In other words, the shape function along

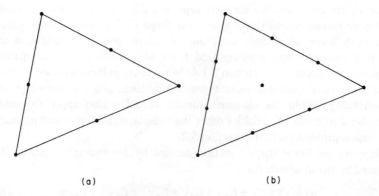

(a)   (b)

Figure 8.2   Higher-order triangular elements: (a) linear strain triangle; (b) quadratic strain triangle

an element interface should be the same for both of the elements concerned. The form of variation along a side of the linear strain triangle shown in figure 8.2a is quadratic, and since there are three nodes along the side this function must be the same for the adjoining elements. Similarly, in the quadratic strain triangle shown in figure 8.2b the cubic variation of displacement or velocity along a side is uniquely defined by values at four nodal points. Both the linear and quadratic strain triangular elements are conforming in the present context.

As an example of a nonconforming element, consider a triangle with three nodes and the following three-parameter shape function

$$w(x, y) = C_1 + C_2 x + C_3 xy \qquad (8.15)$$

The variation of w along a side of the element is in general a quadratic function of position, which is not uniquely defined by the values at the two nodal points on that side. For elements to be conforming, polynomial shape functions should be complete. If terms are omitted, such as the linear term in y in equation 8.15, they should not be of an order less than the highest order present.

The advantage of using elements of a higher order than the CST type is that the shape functions are capable of representing the true variations more accurately. Although the number of elements can therefore be reduced, the reduction in the number of linear equations to be solved may be much less significant because the number of nodal point variables per element is increased. The direct equilibrium formulations described in chapters 3 and 6 are not applicable to variable strain elements. Variational formulations such as those described in sections 3.4 and 6.3 must therefore be used to establish the integrations required over the areas of the elements. The accuracy of a particular finite element solution can be improved by increasing either the number of elements or the order of their shape functions. Since there are no general methods short of trial-and-error for determining which is the more efficient in terms of overall cost, it is often more convenient to increase the number of simple elements.

8.3 Three-dimensional Problems

Finite element methods can in principle be used to solve three-dimensional problems. The three-dimensional equivalent of the triangular element is one in the form of a tetrahedron. A linear shape function similar to equation 3.5 may be used

$$w(x, y, z) = C_1 + C_2 x + C_3 y + C_4 z \qquad (8.16)$$

where the parameters C_1 to C_4 may be found in terms of the values of the variable w at the nodes at the four corners of the element. As with two-dimensional elements, higher-order shape functions can be used.

In practice, the solution of three-dimensional problems is often limited by cost. If sufficient numbers of elements and nodal points are used to give an acceptable degree of accuracy, the resulting set of equations may be prohibitively expensive to solve in terms of both computing time and storage requirements. Suppose an average number of 20 nodal points per co-ordinate direction is required. For a two-dimensional problem the total number of nodes is 400, resulting in 400 equations or pairs of equations for harmonic and biharmonic problems. For a three-dimensional problem this figure is increased to 8000. At the same time the number of nonzero coefficients in the overall stiffness matrix is also increased. For example, using the Gauss–Seidel method discussed in sections 3.6.2 and 3.6.3, the figure of up to 9 nonzero stiffness coefficients per row is increased to 27 for three-dimensional problems. With 8000 nodal points the minimum storage requirements are of the order of a quarter of a million and two million words respectively for harmonic and biharmonic problems. Using an elimination method applied to the banded stiffness matrix as described in section 3.6.1, the bandwidth is similarly increased and the storage requirement is of the order of several million words. Nevertheless, for the reasons given in section 3.6.3, such a large problem would normally be solved by elimination using backing stores.

8.4   Biharmonic Problems Involving Incompressible Materials

In section 6.1.3 it is shown that the finite element method described in chapter 6 for biharmonic problems using displacements as the variables is unsuitable for problems of the plane strain type if the material concerned is incompressible. It is also unsuitable for axi-symmetric biharmonic problems as indicated in section 8.1. Alternative methods must therefore be sought. One simple approach is to employ a value of Poisson's ratio slightly less than the incompressible value of ½. Unfortunately it is generally not possible to use a close enough value without at the same time suffering a severe loss of accuracy.

The formulations of alternative finite element methods for problems involving incompressible materials provide good examples of more general types of approach than those so far examined. Consider, for example, the recirculating viscous flow problem outlined in section 2.2.7. Clearly, a variable other than velocities should be introduced, and the stream function ψ defined by equations 2.51 is convenient. It automatically satisfies the continuity equation 2.20 which expresses the incompressibility condition. Since the velocity components u and v must be at least linear functions of position within an element, the shape function for ψ must be at least quadratic. Now in order to satisfy inter-element compatibility both u and v, and hence the first derivatives of ψ, should be continuous across the element boundaries. One way to encourage such continuity is to use the velocities as nodal point variables in addition to the stream function. A three-node triangular element could be used with ψ, u and v as the unknowns, giving a total of nine nodal point variables. A modified cubic shape function is therefore appropriate, such as

$$\psi(x, y) = C_1 + C_2 x + C_3 y + C_4 x^2 + C_5 y^2 + C_6 xy + C_7 x^3 + C_8 y^3 + C_9 (x^2 y + xy^2)$$

(8.17)

Note that the polynomial is complete and the modification is performed on terms of the highest order present. Since the expressions for u and v obtained by differentiating this function are quadratic, inter-element compatibility is not satisfied with only two nodes per element side.

A slightly different approach is to define a new variable

$$\phi = u + v \qquad (8.18)$$

If ψ and ϕ are used as the unknowns there are only six nodal point variables and the full quadratic shape function involving the first six terms of equation 8.17 can be used. Hence

$$\phi(x, y) = u + v = \frac{\partial \psi}{\partial y} - \frac{\partial \psi}{\partial x}$$

$$= (C_3 - C_2) + (C_6 - 2C_4)x + (2C_5 - C_6)y \qquad (8.19)$$

Although this linear form for ϕ ensures that the sum of the velocities is continuous across inter-element boundaries, it does not necessarily follow that the velocities are individually continuous. While the elements are therefore not strictly conforming, they have been used successfully to solve a range of problems. The constants C_1 to C_6 for a particular element may be obtained in terms of the six nodal point variables. The strain rates are constant over the element and the derivation of the element and overall stiffness matrices is similar to that described in chapter 6. The use of stream function and velocities as variables has considerable advantages for the application of boundary conditions, which as indicated in section 2.2.7 usually involve prescribed values of ψ, u and v.

8.5 Plate and Shell Problems

Many problems involving flat plates or thin curved shells can be solved with the aid of flat plate finite elements. The biharmonic governing differential equation for the deformation of a flat plate is given in section 2.2.8, and in section 1.2.2 an analysis is presented for rigid-jointed structures using elements which are in effect one-dimensional plate elements. Two-dimensional plate elements are generally chosen to be triangular or rectangular in shape. For example, a three-node triangular element could be used, with nodal point variables consisting of the normal displacement w and its two first derivatives with respect to x and y, the local co-ordinates in the plane of the element. With a total of nine variables per element, a modified cubic shape function for w of the form given by equation 8.17 could be used: the present type of problem is analogous to the recirculating viscous flow problem considered in the last section.

8.6   Isoparametric Elements

Although a detailed consideration of finite elements of the isoparametric type is beyond the scope of this book, it would be incomplete without at least a simple introduction to them. A disadvantage of higher-order elements such as those shown in figure 8.2 is that while they have more than two nodes per side, the sides are straight. Therefore, although it may be possible to use a relatively small number of such elements inside the solution domain, a relatively large number may be required near the boundary to adequately fit its shape. The use of elements with curved sides offers an improvement in this respect. Figure 8.3 shows a curved form of the triangular element shown in figure 8.2a. Since there are three nodal points per side of the element the natural forms of function for specifying the shapes of the sides are quadratic functions of position. If the shape function for, say, displacement is of the same order the element is said to be isoparametric, and inter-element compatibility is assured.

8.7   Nonlinear Problems

All the problems so far considered are linear in the sense that they involve the solution of sets of linear algebraic equations. Nonlinearities can be introduced by either geometric or material property effects. Geometric nonlinearities arise, for example, in problems involving solid media in which the strains are sufficiently large to significantly affect the shape of the solution domain. Such problems can be solved by an incremental approach in which the nonlinear analysis is replaced by a series of linear analyses for progressively increasing external loads, after each of which the finite element mesh geometry is recomputed. Alternatively, the overall equations can be solved by an iterative method and the mesh geometry and hence the overall stiffness matrix updated at prescribed intervals during the solution process.

Examples of material nonlinearities include non-newtonian fluids and nonlinearly elastic solids, whose properties are functions of the local state of deformation. Suppose that the viscosity of the fluid involved in the channel flow problem outlined in section 2.2.1 depends on the local rate of deformation in a prescribed manner. Such behaviour can be accommodated in the finite element analysis described in chapter 3 by treating element viscosities not as constants but as functions of the element strain rates, which therefore need to be updated during the Gauss–Seidel solution process. This updating need not necessarily be performed after every cycle of iteration. Direct elimination methods are unsuitable for solving the nonlinear overall equations.

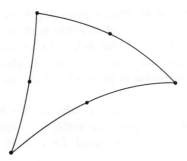

Figure 8.3   A linear strain triangular element with curved sides

Another consequence of nonlinear material behaviour is that the analysis of solid or fluid deformation may become coupled to, say, that of heat transfer. For example, if fluid viscosity in the channel flow problem is a function of the local temperature, the velocity and temperature profiles must be computed simultaneously.

8.8   A Summary of the Finite Element Approach

The various steps involved in the solution of any problem by a finite element method may be summarised as follows.

(1) *Subdivision of the continuum or structure into subregions or finite elements.* Such elements can be one-dimensional (section 1.2.2), two-dimensional (sections 3.1 and 6.1) or three-dimensional (section 8.3) according to the type of problem, and each one is an essentially simple unit. Some of the criteria for choosing the size, shape and distribution of elements are discussed in section 4.1.

(2) *Selection of the nodal point variables and shape functions.* In most of the problems considered here nodal point displacements or velocities are used as the variables. Sections 5.2 and 8.4 provide examples of the use of a stress function and stream function respectively, while sections 1.2.2, 8.4 and 8.5 describe methods which involve the use of derivatives at the nodal points. Shape functions are usually polynomials, the simplest of which are linear such as those introduced in sections 3.1.1, 6.1.1 and 8.3. The use of higher-order shape functions is described and discussed in sections 1.2.2, 8.2 and 8.4.

(3) *Derivation of the element behaviour.* A relationship between, for example, nodal point forces and displacements is obtained for a typical element, by either a direct equilibrium analysis (sections 1.2.2, 3.1 and 6.1) or a variational method (sections 3.4, 6.3 and 8.1). The behaviour of all the individual elements may be computed with the aid of this relationship.

(4) *Assembly of the algebraic equations for the overall continuum or structure.* The characteristics, such as stiffness, of the individual elements are added together to determine the behaviour of the overall system. The resulting algebraic equations are often linear (sections 1.2.2, 3.2 and 6.2), although material properties or geometric effects may make them nonlinear (section 8.7).

(5) *Application of the boundary conditions.* The boundary conditions or restraints associated with a particular problem are applied by modifying the overall algebraic equations as described, for example, in sections 1.2.2, 3.5 and 6.5.

(6) *Solution of the overall equations.* Direct and iterative methods for solving linear overall equations are considered in sections 3.6 and 6.4, and appendixes A and B, and the modifications necessary for solving nonlinear problems are discussed in section 8.7.

(7) *Computation of further results.* In addition to the values of the nodal point variables obtained from the overall equations, other results such as integrals over the solution domain (sections 5.1.3 and 5.2.3) or element and nodal point stresses (section 6.6.7) may also be required.

8.9　Concluding Remarks

The advent of high-speed digital computers and the development of numerical methods such as those of the finite element type have broadened considerably the scope of analysis in many branches of engineering. Finite element methods are in many ways very similar to the older finite difference type but offer some advantages. The flexibility of element size, shape and distribution makes it possible to fit complicated solution domain boundary shapes and to choose nonuniform internal distributions of nodal points to suit the problem concerned. It is often unnecessary to resort to the sophisticated co-ordinate systems sometimes employed in finite difference methods in order to fit particular boundary shapes. Derivative boundary conditions are usually easier to handle using finite element methods, as are inhomogeneous solution domains involving abrupt changes of material properties.

Finite element methods are powerful tools available to the designer of engineering components and systems. Associated with this power, however, is a relatively high cost in terms of computer time and storage capacity. A substantial amount of engineering judgement is required to decide whether a particular problem is worth subjecting to detailed analysis and, if it is, to choose an appropriate distribution and fineness for the mesh of elements. As with any numerical method employing a digital computer, a great deal of care must be taken to ensure that the program used has been thoroughly tested and that the data are supplied to it in the correct form. The results obtained should be interpreted with caution, bearing in mind the assumptions and limitations of the particular method used.

Appendix A Gaussian Elimination

Gaussian elimination is a direct method for solving sets of simultaneous linear algebraic equations of the general form

$$
a_{11}x_1 + a_{12}x_2 + \ldots + a_{1n}x_n = b_1
$$
$$
a_{21}x_1 + a_{22}x_2 + \ldots + a_{2n}x_n = b_2
$$

$$\tag{A.1}$$

$$
a_{n1}x_1 + a_{n2}x_2 + \ldots + a_{nn}x_n = b_n
$$

where $x_1, x_2, \ldots, x_n$ are the unknowns, and the coefficients a_{ij} and b_i are all known constants. This set of equations can be expressed in matrix form as follows

$$
\begin{bmatrix}
a_{11} & a_{12} \ldots a_{1n} \\
a_{21} & a_{22} \ldots a_{2n} \\
\vdots & \vdots \\
a_{n1} & a_{n2} \ldots a_{nn}
\end{bmatrix}
\begin{bmatrix}
x_1 \\
x_2 \\
\vdots \\
x_n
\end{bmatrix}
=
\begin{bmatrix}
b_1 \\
b_2 \\
\vdots \\
b_n
\end{bmatrix}
\tag{A.2}
$$

$$
AX = B \tag{A.3}
$$

Both gaussian elimination and other methods of solution are considered in more detail by Fenner (1974).

The unknowns are successively eliminated by algebraic manipulation. The first equation can be used to eliminate x_1 from the remaining $n - 1$ equations. The modified second equation is then used to eliminate x_2 from the remaining $n - 2$ equations, and so on until the last equation contains only x_n. Thus x_n may be found, followed by all the other unknowns, by back substitution. Let the coefficients shown in equations A.1 and A.2 be given the notation $a_{ij}^{(1)}$, $b_i^{(1)}$.

After the kth elimination, the modified coefficients are

$$a_{ij}^{(k+1)} = a_{ij}^{(k)} - \phi a_{kj}^{(k)}$$
$$b_i^{(k+1)} = b_i^{(k)} - \phi b_i^{(k)}$$

(A.4)

where $\phi = a_{ik}^{(k)}/a_{kk}^{(k)}$; $i = k + 1$, $k + 2$, ..., n and $j = k$, $k + 1$, ..., n. Note that the vector B is treated just like a column of A, and advantage can be taken of this fact to simplify the computer programming of the process. The final set of equations is

$$a_{11}^{(1)}x_1 + a_{12}^{(1)}x_2 + \ldots + a_{1n}^{(1)}x_n = b_1^{(1)}$$
$$a_{22}^{(2)}x_2 + \ldots + a_{2n}^{(2)}x_n = b_2^{(2)}$$
$$\vdots \qquad \qquad \vdots$$
$$a_{nn}^{(n)}x_n = b_n^{(n)}$$

(A.5)

Expressed in matrix terminology, the elimination process triangularises A. The unknowns are obtained in reverse order

$$x_n = b_n^{(n)}/a_{nn}^{(n)}$$

(A.6)

$$x_i = \left(b_i^{(i)} - \sum_{j=i+1}^{n} a_{ij}^{(i)}x_j\right)\bigg/a_{ii}^{(i)}$$

(A.7)

where $i = n - 1$, $n - 2$, ..., 1. As the elimination process does not affect the value of the determinant of A, it may be found from the triangularised matrix as

$$|A| = a_{11}^{(1)}a_{22}^{(2)}a_{33}^{(3)} \ldots a_{nn}^{(n)}$$

(A.8)

If the value of this determinant is very small or zero, the equations are said to be ill-conditioned or singular. Singular sets of equations do not have unique solutions, and solutions obtained from ill-conditioned sets may be subject to significant errors (Fenner (1974)).

In order to test for ill-conditioning it is convenient to determine the mean magnitude of the coefficients in A

$$\bar{a} = \frac{1}{n^2} \sum_{i=1}^{n} \sum_{j=1}^{n} |a_{ij}|$$

(A.9)

Since $|A|$ involves the product of n coefficients it is appropriate to compare its magnitude with that of $\bar{a}^n$, by means of a ratio

$$R = |A|/\bar{a}^n$$

(A.10)

and the equations are ill-conditioned if R is small compared with one.

A great many arithmetic operations are involved in solving large sets of equations by elimination. Any errors introduced, such as roundoff errors (see section 1.1), tend to be magnified and may become unacceptably large.

```
      SUBROUTINE  ELIMIN(A,X,MEQN,NROW,NCOL,DET,RATIO)
C
C  SUBPROGRAM FOR SOLVING SIMULTANEOUS LINEAR EQUATIONS BY GAUSSIAN
C  ELIMINATION WITH PARTIAL PIVOTING.
C
      DIMENSION  A(NROW,NCOL),X(NROW)
      NEQN=MEQN
      IF(NEQN.LE.NROW.AND.NEQN.LE.NCOL-1) GO TO 1
      WRITE(6,61)
  61  FORMAT(33HOSTOP - DIMENSION ERROR IN ELIMIN)
      STOP
C
C  FIND MEAN COEFFICIENT MAGNITUDE.
  1   AMEAN=0.
      DO 2 I=1,NEQN
      DO 2 J=1,NEQN
  2   AMEAN=AMEAN+ABS(A(I,J))
      AMEAN=AMEAN/FLOAT(NEQN*NEQN)
C
C  COMMENCE ELIMINATION PROCESS.
      JMAX=NEQN+1
      NEQNM1=NEQN-1
      DO 6 IEQN=1,NEQNM1
C
C  SEARCH LEADING COLUMN OF THE COEFFICIENT MATRIX FROM THE DIAGONAL
C  DOWNWARDS FOR THE LARGEST ELEMENT AND MAKE THIS THE PIVOTAL ELEMENT.
      IMIN=IEQN+1
      IMAX=IEQN
      DO 3 I=IMIN,NEQN
  3   IF(ABS(A(I,IEQN)).GT.ABS(A(IMAX,IEQN))) IMAX=I
      IF(IMAX.EQ.IEQN) GO TO 5
      DO 4 J=IEQN,JMAX
      AA=A(IEQN,J)
      A(IEQN,J)=A(IMAX,J)
  4   A(IMAX,J)=AA
C
C  ELIMINATE X(IEQN) FROM EQUATIONS (IEQN+1) TO NEQN, FIRST TESTING FOR
C  NONZERO PIVOTAL ELEMENT.
  5   IF(ABS(A(IEQN,IEQN)/AMEAN).LT.1.E-8) GO TO 10
      DO 6 I=IMIN,NEQN
      FACT=A(I,IEQN)/A(IEQN,IEQN)
      DO 6 J=IMIN,JMAX
  6   A(I,J)=A(I,J)-FACT*A(IEQN,J)
C
C  SOLVE THE UPPER-TRIANGULAR SET OF EQUATIONS BY BACK SUBSTITUTION.
      IF(ABS(A(NEQN,NEQN)/AMEAN).LT.1.E-8) GO TO 10
      X(NEQN)=A(NEQN,JMAX)/A(NEQN,NEQN)
      DO 8 L=2,NEQN
      I=NEQN+1-L
      SUM=A(I,JMAX)
      IP1=I+1
      DO 7 J=IP1,NEQN
  7   SUM=SUM-A(I,J)*X(J)
  8   X(I)=SUM/A(I,I)
C
C  EVALUATE DETERMINANT OF COEFFICIENT MATRIX AND COMPARE WITH
C  ORIGINAL COEFFICIENTS.
      DETA=1.
      DO 9 I=1,NEQN
  9   DETA=DETA*A(I,I)
      DET=DETA
      RATIO=DETA/AMEAN**NEQN
      RETURN
  10  DET=0.
      RETURN
      END
```

Figure A.1   Subprogram for the gaussian elimination method

Equations A.4 show that the elimination process involves many multiplications by the factors ϕ. In order to minimise the effect of any errors in the coefficients $a_{kj}^{(k)}$ and $b_k^{(k)}$, these factors should be as small as possible, and certainly less than one. Thus, the 'pivotal' coefficient $a_{kk}^{(k)}$ should be the largest one in the leading column of the remaining submatrix

$$| a_{kk}^{(k)} | > | a_{ik}^{(k)} | \quad i = k + 1, k + 2, \ldots, n \tag{A.11}$$

This condition also serves to avoid division by zero in equations A.4, and can be achieved by interchanging equations, a technique known as partial pivoting.

Figure A.1 shows a subprogram named ELIMIN for implementing the gaussian elimination method. A detailed description and flow chart are given by Fenner (1974). The arguments include the array A of coefficients and the solution vector X. The variables NROW and NCOL, which enter the maximum numbers of rows and columns permitted in A, are used for dimensioning purposes. Normally, NCOL=NROW+1 to allow A to store the coefficients of vector B (as the $(n + 1)$th column of the matrix A). The argument MEQN enters the actual number of equations to be solved, n, and in order to minimise the execution time its value is assigned to the local variable NEQN. The remaining arguments DET and RATIO return the values of $|A|$ and the ratio R to the calling program.

After the acceptability of the number of equations has been tested, the mean coefficient magnitude is computed according to equation A.9, and stored in the variable AMEAN. The elimination process is started by first defining JMAX as the number of coefficients in each row of the extended matrix, and NEQNM1 as the number of eliminations to be performed. Then each equation, with the exception of the last, is used in turn to eliminate the corresponding unknown. The current equation number is given by IEQN, and is equivalent to k used above. Before performing the necessary eliminations with a particular equation, however, a search is made down the leading column of the remaining submatrix for the largest coefficient, to satisfy equation A.11. The search technique locates the row number of this largest coefficient, IMAX. If the existing pivotal coefficient is not the largest, then the rows are interchanged.

Despite the search for the largest coefficient, the resulting pivotal coefficient is still extremely small or zero if the equations are very ill-conditioned or singular. The following test is made for its relative magnitude

$$| a_{kk}/\bar{a} | < 10^{-8} \tag{A.12}$$

and the problem is rejected if this condition is satisfied. Rejection is indicated by setting DET to zero, which may be detected by the calling program. If the partial pivoting is successful, however, the eliminations defined by equations A.4 are performed, with the variable FACT being used to store the values of the factor ϕ.

After testing the magnitude of the last diagonal coefficient, the back substitutions defined in equations A.6 and A.7 are performed to find the required solutions. Finally, the value of $|A|$ is obtained according to equation A.8, using the local variable DETA to accumulate the value of the required product, which is then assigned to the argument DET. The ratio R is computed according to equation A.10 and its value stored in the argument RATIO.

Appendix B The Gauss-Seidel Method

The Gauss–Seidel method is an iterative technique for solving sets of simultaneous linear algebraic equations of the general form displayed in equations A.1, appendix A. It involves expressing each unknown as a function of the others, as follows

$$x_i^{(m)} = \frac{1}{a_{ii}} \left(b_i - \sum_{j=1}^{i-1} a_{ij}x_j^{(m)} - \sum_{j=i+1}^{n} a_{ij}x_j^{(m-1)} \right) \qquad (B.1)$$

where the superscripts denote iteration numbers. Note that the most up-to-date values of the unknowns are used. Clearly, it is essential for all the diagonal coefficients a_{ii} to be nonzero, and for the process to be convergent even more stringent conditions should be satisfied.

In order to test for convergence as the number of iterations is increased, the changes in the unknowns between successive iterations can be compared with their current values. An appropriate criterion is

$$e_r = \sum_{i=1}^{n} |\Delta x_i| \Big/ \sum_{i=1}^{n} |x_i^{(m)}| < \alpha \qquad (B.2)$$

where e_r is the relative error, $\Delta x_i = x_i^{(m)} - x_i^{(m-1)}$ and α is a suitably small tolerance.

The choice of starting values for the unknowns does not normally affect whether the Gauss–Seidel process converges, and often has comparatively little effect on the number of iterations required. It is possible to predict whether convergence is likely to be achieved with a particular set of equations. Varga (1962) has stated the sufficient condition for convergence as that of 'diagonal dominance' of the coefficient matrix A. If A is diagonally dominant, then

$$|a_{ii}| \geqslant \sum_{\substack{j=1 \\ j \neq i}}^{n} |a_{ij}| \quad \text{for } i = 1, 2, \ldots, n \qquad (B.3)$$

and the inequality is satisfied for at least one row. While diagonal dominance is sufficient to ensure convergence, it may not be necessary, provided these conditions are only mildly contravened.

It is often possible to improve the rate of convergence by a technique which is generally known as over-relaxation. Equation B.1 provides new estimates, $x_i^{(m)}$, which, provided the process is convergent, are closer to the required solutions than the $x_i^{(m-1)}$. Over-relaxation applies a limited amount of extrapolation from these two sets of estimates towards the final solutions. Thus, if $\tilde{x}_i^{(m)}$ are the values obtained from equations B.1, the extrapolated values after the mth iteration are

$$x_i^{(m)} = x_i^{(m-1)} + \omega(\tilde{x}_i^{(m)} - x_i^{(m-1)}) \qquad \text{(B.4)}$$

where ω is an over-relaxation factor, which is the same for all the equations. For a particular set of linear equations there is an optimum value of ω, normally in the range $1 < \omega < 2$. The purpose of over-relaxation is to accelerate convergence, rather than to promote convergence in an otherwise divergent iteration scheme. The use of too large a value of ω can cause divergence.

For computer programming purposes, it is convenient to rewrite equations B.4 and B.1 with the aid of the changes in the unknowns, Δx_i, introduced in equation B.2. Thus

$$x_i^{(m)} = x_i^{(m-1)} + \omega \Delta x_i \qquad \text{(B.5)}$$

$$\Delta x_i = \frac{1}{a_{ii}} \left(b_i - \sum_{j=1}^{n} a_{ij} x_j \right) \qquad \text{(B.6)}$$

where the latest values of the unknowns are used in the summations.

Unfortunately, it is not a simple matter to predict the optimum value of ω (see Varga (1962) and Isaacson and Keller (1966)), and the usual approach is an empirical one. The required value can be found as that which gives either convergence to a particular tolerance with the minimum number of iterations, or the minimum error after a certain number of iterations. The optimum over-relaxation factor is determined by the number of equations and the nature of the coefficient matrix A. Further discussion and practical applications of the Gauss—Seidel method are provided by Fenner (1974).

Bibliography

Bird, R. B., Stewart, W. E., and Lightfoot, E. N. (1960), *Transport Phenomena*, Wiley, New York

Crandall, S. H. (1956), *Engineering Analysis*, McGraw-Hill, New York

Desai, C. S., and Abel, J. F. (1972), *Introduction to the Finite Element Method*, Van Nostrand Reinhold, New York

Fenner, Roger T. (1974), *Computing for Engineers*, Macmillan, London

Ford, H. (1963), *Advanced Mechanics of Materials*, Longmans, London

Gosman, A. D., Pun, W. M., Runchal, A. K., Spalding, D. B., and Wolfshtein, M. (1969), *Heat and Mass Transfer in Recirculating Flows*, Academic Press, London

Isaacson, E., and Keller, H. B. (1966), *Analysis of Numerical Methods*, Wiley, New York

McCracken, D. D. (1972), *A Guide to Fortran IV Programming*, 2nd edn, Wiley, New York

Nath, B. (1974), *Fundamentals of Finite Elements for Engineers*, Athlone Press, London

Schechter, R. S. (1967), *The Variational Method in Engineering*, McGraw-Hill, New York

Varga, R. S. (1962), *Matrix Iterative Analysis*, Prentice-Hall, London

Zienkiewicz, O. C. (1971), *The Finite Element Method in Engineering Science*, McGraw-Hill, London

Index